MANIPULATION OF GROWTH IN FARM ANIMALS

CURRENT TOPICS IN VETERINARY MEDICINE AND ANIMAL SCIENCE

MANIPULATION OF GROWTH IN FARM ANIMALS

A Seminar in the CEC Programme of Coordination
of Research on Beef Production, held in Brussels,
December 13–14, 1982

Sponsored by the Commission of the European
Communities, Directorate-General for Agriculture,
Coordination of Agricultural Research

Edited by

J.F. Roche and D. O'Callaghan
University College Dublin
Veterinary Field Station
Ballycoolin Road
Finglas, Dublin 11
Ireland

1984 **MARTINUS NIJHOFF PUBLISHERS**
a member of the KLUWER ACADEMIC PUBLISHERS·GROUP
BOSTON / THE HAGUE / DORDRECHT / LANCASTER
for
THE COMMISSION OF THE EUROPEAN COMMUNITIES

Distributors

for the United States and Canada: Kluwer Boston, Inc., 190 Old Derby Street, Hingham, MA 02043, USA
for all other countries: Kluwer Academic Publishers Group, Distribution Center, P.O.Box 322, 3300 AH Dordrecht, The Netherlands

Library of Congress Cataloging in Publication Data

```
Main entry under title:

Manipulation of growth in farm animals

    Current topics in veterinary medicine and animal
science)
    "Sponsored by the Commission of the European
Communities, Directorate-General for Agriculture,
Coordination of Agricultural Research."
    "Publication arranged by: Commission of the
European Communities, Directorate-General, Information
Market, and Innovation"--T.p. verso.
    1. Livestock--Growth--Congresses.  2. Anabolic
steroids in animal nutrition--Congresses.  I. Roche, J. F.
II. O'Callaghan, D.  III. Commission of the European
Communities.  Coordination of Agricultural Research.
IV. Commission of the European Communities.  Directorate
General, Information, Market, and Innovation.  V. Series.
SF768.M34  1984      636.08'5      83-23713
```
ISBN 978-94-010-9484-9   ISBN 978-94-010-9482-5 (eBook)
DOI 10.1007/978-94-010-9482-5

EUR 8919 EN

Book information

Publication arranged by: Commission of the European Communities, Directorate-General Information Market and Innovation, Luxembourg

Copyright/legal notice

CONTENTS

P R E F A C E
‾‾‾‾‾‾‾‾‾‾

This publication contains the proceedings of a
workshop held at C.E.C. headquarters, Brussels, December
13th and 14th 1982, under the auspices of the Commission
of the European Communities, as part of the EEC programme
on co-ordination of research in beef production. The
aim of the workshop was to update the existing position
as regards use, safety, metabolism and residues of
anabolic agents. The endocrine control of growth was
discussed and alternative methods to increase growth
were outlined. The organiser of the workshop was
Dr. J.F. Roche (IRL) in conjunction with Mr. J. Connell
of C.E.C.

Thanks are due to all the people who helped to make
this a very successful workshop. The sessions were
chaired by Professor B. Hoffmann, Professor H. Karg,
Dr. R.J. Heitzman and Dr. J.F. Roche.

RECENT STUDIES ON PHARMACOKINETICS AND RESIDUES OF ANABOLIC AGENTS IN BEEF CATTLE AND OTHER FARM ANIMALS

R. J. Heitzman, A. Carter, S.N. Dixon, D.J. Harwood and M. Phillips

Agricultural Research Council,
Institute for Research on Animal Diseases,
Compton, Newbury, Berkshire, U.K., RG16 ONN.

ABSTRACT

Research on measurement of residues and pharmacokinetics of zeranol, trenbolone, oestradiol, testosterone and progesterone and also the stilbenes DES and hexoestrol has progressed. New data on residue concentrations of anabolic agents are listed and the introduction of new developments in assay techniques described. Radioimmunoassay (RIA) is still the most widely used method for residue measurement and there have been new advances in the production of both polyclonal and monoclonal antibodies for zeranol, trenbolone and 17α-OH trenbolone, the major metabolite of trenbolone.

Data on the absorption and half-lives of endogenous steroids from compressed pellets is reviewed. Investigations are proceeding into the metabolism of trenbolone and zeranol by the liver and their clearance into the bile of ruminants is being measured. The metabolism of trenbolone in ruminants was shown to be substantially different from that seen in rats, pigs, guinea pigs, rabbits and humans.

INTRODUCTION

Techniques for the measurement of residues of anabolic agents in farm animals, their meat and their meat products have advanced during the last 7 years due to the introduction of new methodology. Originally the thin-layer chromatographic method of Verbeke (1979) was used in many laboratories but this method has been largely superceded by more rapid and sensitive radioimmunoassay (RIA) methods (Dixon & Heitzman, 1981; Karg & Vogt, 1981; Agthe, 1980; Hoffmann, 1978). This paper discusses the recent progress in methods of residue analysis and presents some new data on residues of the "permitted five substances", oestradiol-17β, testosterone, progesterone, zeranol and trenbolone and also the "banned substances", stilbenes, hexoestrol and diethylstilboestrol (DES).

A major limitation of all the analythical methods for measuring residues is the time consuming preliminary extraction procedures prior to final analysis. It is now possible to separate and remove many of the unwanted interfering substances by the use of automated high performance liquid chromatography (HPLC). The combination of HPLC and RIA for residue measurement is discussed.

The variable quality and limited supply of antisera for RIA has restricted the widespread use of RIA. However, new techniques in biotechnology have enabled monoclonal antibodies to be produced for the assay of anabolic agents. These antibodies have increased the specificity of assays and may be produced in relatively large amounts. Some initial studies with monoclonal antibodies to zeranol are reported.

There has been little new information on the metabolism and pharmacokinetics of anabolic agents. Further knowledge is needed on the metabolism of the xenobiotic anabolic agents and in particular the role of the liver in detoxification and clearance of these substances. Some initial studies into the pharmacokinetics of trenbolone acetate administered to cattle are discussed. The metabolism of trenbolone acetate is different in cattle and rats (Pottier et al., 1981) and this paper compares the metabolism in several other species.

The development of better delivery systems for sustaining a nearly constant absorption of anabolic agents from a depot site into the peripheral circulation is necessary. The introduction of implants of silastic material impregnated with oestradiol-17β has been a major development (Wagner & Pankhurst, 1981). There is also a sustained and gradual absorption of oestradiol-17β from implants containing a mixture of trenbolone acetate and oestradiol-17β (Riis & Suresh, 1976; Heitzman & Harwood, 1977; Harrison, 1981). Other commercially available implants contain mixtures of oestradiol-17β with either testosterone or progesterone. Experiments have been carried out to examine the hypothesis that the absorption of oestradiol-17β from implants of compressed pellets is influenced by mixing with a second steroid.

ZERANOL

Assay Using Polyclonal and Monoclonal Antibodies

Dixon (1980) developed a RIA method for measuring
zeranol which was later used to measure residues of zeranol
in urine of cattle and sheep. However, the antiserum was
raised against zeranol-7α-hemisuccinate-bovine serum albumin
(Z-7H-BSA) and was found to cross react (48%) with the mycooestrogen zearalenone (Dixon & Russell, unpublished data).
Zearalenone may be a contaminant of cereal grain. If such
grain is used in animal feeds residues may persist in tissues
of animals which have been fed on contaminated feedstuffs.
It was therefore important to develop a RIA method with
greater specificity for zeranol.

This was achieved by synthesising an antigen based upon
the substitution at the 16 position rather than at the 7
position. Zeranol-16-carboxy-propyl-ether-human serum albumin
(Z-CPE-HSA) was synthesised and used to produce an antiserum
in sheep. The same antigen was also injected into mice and
3 days later the spleen cells were collected. These cells
were fused with mouse myeloma cells to form hybridoma cells
which were screened and selected for the production of monoclonal antibodies against zeranol. Monoclonal antibodies were
also produced against the antigen Z-7H-BSA. The cross-
reactivities of the polyclonal and monoclonal antibodies were
measured and the results are shown in Table 1.

TABLE 1. Cross reactions of zeranol antibodies

Antigen	Z-7-hemisuccinate-BSA		Z-16-carboxypropyl-ether-HSA	
Type antibody	Polyclonal	Monoclonal	Polyclonal	Monoclonal
Compound		Cross reactions %		
Zeranol	100	100	100	100
Zearalenone	48	17	<0.01	<0.01
Taleranol	26	12	2	<0.01
α-zearalenol	71	17	30	13
β-zearalenol	9	<0.01	<0.01	<0.01
Zearalanone	100	100	<0.01	<0.01

Zeranol is 7α-zearalanol, taleranol is 7β-zearalanol,
zearalenone is a mycooestrogen.

4

The antibodies raised against the antigen Z-CPE-HSA reduced cross-reactivity and the polyclonal antibody was used in a RIA to measure the residues of zeranol in urine of sheep treated with zeranol (Dixon & Russell, unpublished data). The results are shown in Figure 1 for both antisera and indicate the presence of residues throughout and beyond the recommended withdrawal period of 40 days.

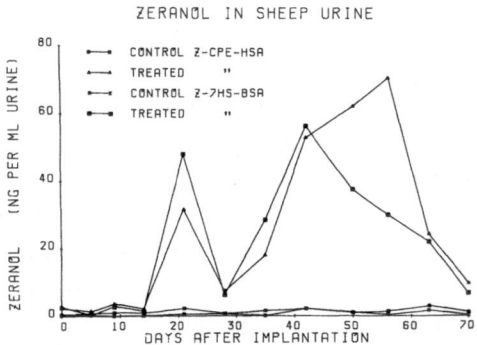

Fig. 1.   Six sheep were implanted with 12 mg zeranol.

The monoclonal antibodies are now being tested in RIA systems for residue measurement in bile, urine and faeces of farm animals.   Initial results are encouraging because their cross-reaction with non-specific interfering substances found in these materials is much less.

A further great advantage of monoclonal antibody production is that a continuous supply of antibodies with identical characteristics may be produced repeatedly whereas with in vivo production systems the spectrum of antibody populations vary.

TRENBOLONE ACETATE
Pharmacokinetics

Following the i.v. injection of 2.8 g or 5 g ^{3}H-trenbolone acetate into cattle >80% of the radiolabel was collected in the bile in the following 8 hours.   The major metabolite in the bile was 17α-trenbolone (Pottier et al.,

1981; Heitzman, unpublished data).

In a further study in one steer 300 mg trenbolone acetate was injected into a jugular vein and bile, blood and urine were collected over an 8 hour period. The concentrations of α-trenbolone in bile and urine and α- and β-trenbolone in plasma are shown in Figure 2. There was very rapid clearance of the drug from the plasma into the bile and urine.

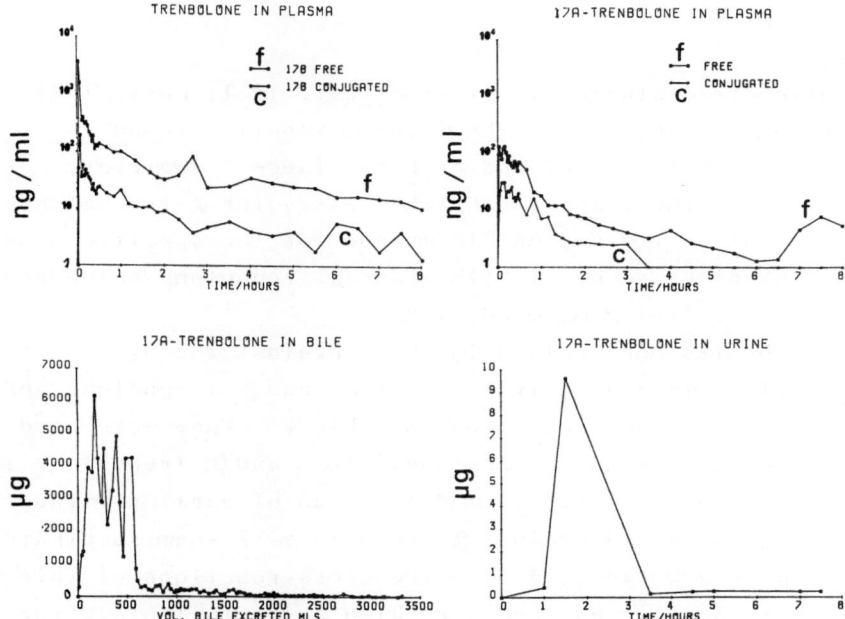

Figure 2. A steer was injected intravenously with 300 mg trenbolone acetate at time 0. Samples were analysed by RIA. The concentrations of 17α-trenbolone in bile and urine are the sum of both free and conjugated steroid.

We suggest that in ruminants trenbolone acetate is metabolised by the following route. Trenbolone acetate enters the bloodstream where it is rapidly hydrolysed to 17β-hydroxy-trenbolone (β-trenbolone), the biologically active anabolic agent. After uptake by the liver it is rapidly converted to the 17-keto derivative, trendione. The trendione reenters the circulation in ruminants where it is rapidly converted to 17α-hydroxy-trenbolone (α-trenbolone). We have been unable to confirm the suggestion of Pottier <u>et al</u>. (1981) and Rico (1982) who claimed that α-trenbolone was formed in the liver cell. The enzymic conversion to α-trenbolone occurred only with systems prepared from whole blood of cattle, sheep and goats and not with pigs, humans, rodents, guinea pigs and rabbits. The α-trenbolone is eventually conjugated in liver and excreted into the bile as the main biliary metabolite of trenbolone acetate.

<u>Residues</u>

Radiometric studies (Pottier <u>et al</u>., 1981; Ross, 1981) have shown α-trenbolone is the major metabolite found in liver, kidney bile, urine and probably faeces. However, until recently there was no suitable assay for α-trenbolone in any tissue or fluid. An RIA method for the specific measurement of residues of both α and β-trenbolone has been developed (Phillips & Harwood, 1982).
Serum antibodies were raised in sheep against the 3-carboxy-methoxime-BSA derivatives of α- and β-trenbolone and their cross-reactions are shown in Table 2. They were used for the RIA measurement of residues of α- and β-trenbolone in tissues and fluids by simple modification of earlier methods using antisera raised against β-trenbolone-17 -hemisuccinate-BSA (Hoffmann & Oettel, 1975). The cross-reactions of this latter less specific antibody are also shown in Table 2 for comparison.

The initial results for residues in edible tissues agree with the radiometric data, i.e. the concentrations of α-trenbolone in liver, kidney, bile and urine are many times higher than those of β-trenbolone, whereas there are greater concentrations of β-trenbolone in muscle and plasma. Also the results clearly show (see Table 3) that the method using the less specific antibody does not give an accurate estimation of residues of α- and β-trenbolone.

Monoclonal antibodies are now being produced against both α and β-trenbolone and will be examined in comparison with the polycolonal antibodies.

TABLE 2. Cross-reactions of trenbolone antiserum

Antigen	17β-OH-TBOH 3CMO-BSA	17α-OH-TBOH 3CMO-BSA	TBOH-17β- hemisuccinate-BSA
Species	Sheep M3	Sheep M336	Rabbit
Compound			
Trenbolone acetate	<0.001	<0.001	100
17β-OH trenbolone	100	<0.001	100
17α-OH trenbolone	<0.001	100	10
Trendione	<0.001	<0.001	50
Testosterone	<0.001	<0.001	<0.5

TBOH is trenbolone, BSA is bovine serum albumin, 3CMO is
3-carboxy-methoxime.

TABLE 3. Tissue residues of 17α and 17β-OH trenbolone
from a cow treated with trenbolone acetate (300 mg)

		Trenbolone (μg/kg fresh tissue)		
		Specific 17β assay	Specific 17α assay	17 aspecific assay
Muscle (diaphragm)	Control	0.00	0.05	0.02
	Treated	0.27	0.04	0.08
Liver	Control	0.11	0.22	0.14
	Treated	0.28	1.42	0.74
Kidney	Control	0.03	0.07	0.11
	Treated	0.16	0.41	0.28
Fat	Control	0.02	0.03	0.06
	Treated	0.25	0.15	0.21

STILBENES

Results

There has been great activity to improve the methods for
measuring stilbene residues in animals and edible tissues.
There is still a need for a rapid, reliable and cheap method
for routine measurement of residues, particularly since the
directive (81/602/EEC) prohibiting the use of stilbenes in
farm animals. An additional method is frequently necessary
to confirm the identification of stilbenes.

RIA is the preferred method for initial screening of samples especially bile, urine and faeces. Preliminary extraction procedures are simple for bile (Harwood, Heitzman & Jouquey, 1981) and urine (Hoffmann & Laschutza, 1980) and slightly more complex for faeces (Heitzman & Harwood, 1982). However, Stephany (1982) reported that an additional chromatographic step was necessary before a reliable RIA could be performed on urine. The introduciton of automated high performance liquid chromatography (HPLC) for purifying tissue extracts is most promising and has been successfully used for the assay of DES in urine of cattle (Stephany, 1982). It is reported that 50-100 samples can be purified on HPLC every 24 h and this rapid throughput may offset the high capital cost of HPLC.

In the United Kingdom hexoestrol was used in poultry, sheep and cattle before the ban. Residues of hexoestrol in cattle (Heitzman & Harwood, 1982), sheep (Harwood, Heitzman & Jouquey, 1980) and poultry (Herriman et al., 1982) have been measured.

Chickens were caponised with 12 mg hexoestrol, thus it was not surprising, although somewhat disturbing, to report much higher concentrations of residues in treated chickens than in sheep and cattle (see Table 4). Every tissue examined from caponised birds contained residues and thus the RIA could be used with confidence to detect the use of stilbenes in poultry.

TABLE 4. Residues of hexoestrol (μg/kg fresh tissue)

Tissue	Cattle	Sheep	Poultry
Muscle	0.03	0.08	0.5
Liver	0.1	1.0	6.2
Kidney	0.5	3.0	-
Fat	-	<0.05	1.27

Residues following administration of implants of 36 or 45 mg hexoestrol to bulls and steers were still present in tissues, especially kidney, urine, bile and faeces in the majority of samples collected 90-153 days after implantation (Heitzman & Harwood, 1982). The results in Table 5 show that confirmation of treatment with hexoestrol is possible for at least 100 days after implantation.

TABLE 5. Residues of hexoestrol in bulls and steers[a]

Days after implantation	% samples containing hexoestrol					
	Faeces	Bile	Urine	Liver	Kidney	Muscle
90-104	100	100	89	100	100	22
111-153	69	69	62	54	69	38

[a]Data from Heitzman & Harwood (1982)

The search for alternative methods of measuring stilbenes continues. One possible development is the improvement of an on-line detection system for HPLC. At present the sensitivity of detectors is not adequate for residues < 1 μg/kg tissue although an electrochemical detection system has been used with some success for DES and also HPLC coupled with mass-spectrometry (see Stephany, 1982).

ENDOGENOUS STEROIDS
Pharmacokinetics: Absorption and Half-Life of Steroids in Combined Preparations

A constant delivery of the minimum amount of anabolic agent necessary for the maximum anabolic response is desirable and also it reduces the risk of unwanted residues and the wasteful use of agent. Oestradiol-17β is a potent anabolic agent in veal calves, steers and wethers when the drug is delivered over a long period of time. Growth promotion is poor following implantation of compressed pellets containing only oestradiol-17β because they are absorbed too rapidly (Heitzman et al., 1981). However, it has now been shown that the delivery of oestradiol-17β from implants may be maintained at nearly constant rates over long periods by either a) implanting the oestradiol-17β in compressed pellets containing an intimate mixture of oestradiol-17β with 7-10 times its own weight of trenbolone acetate,

10

testosterone or progesterone (Harrison, 1981) or b) incorporating oestradiol-17β into an inert but permeable silastic
polymer implant (Wagner & Pankhurst, 1981).

A study (Harrison, 1981) has been carried out in wether
sheep using combined preparations of oestradiol-17β and
either trenbolone acetate, testosterone or progesterone. The
concentrations of the four steroids were measured by RIA in
plasma taken from both jugular veins. Similar measurements
were made on sheep implanted with either a) oestradiol-17β
b) separate formulations of oestradiol-17β and trenbolone
acetate (TBA/OE$_1$) c) placebo (control group). Figure 3 gives
details of the treatments and shows the plasma concentrations
of oestradiol-17β in the vein ipsilateral to the implant.

The concentrations of oestradiol-17β in sheep treated
with combined implants remained significanly higher than that
of controls for 15-16 weeks after implantation. The
concentrations of oestradiol-17β in plasma of sheep implanted
with oestradiol-17β either alone or placed separately from
implants of trenbolone acetate were initially higher than the
concentration in controls, but were the same as control
values 8-16 weeks after implantation.

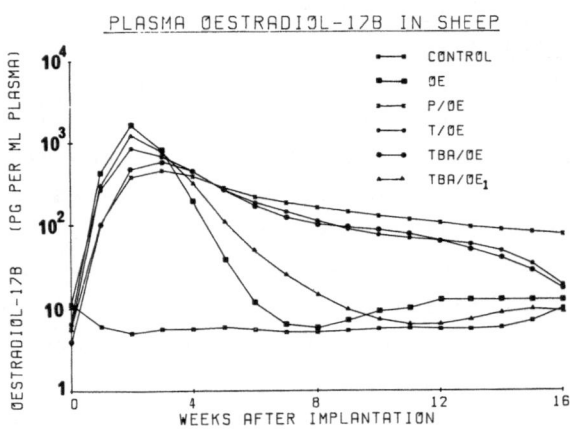

Fig. 3. Concentrations of oestradiol-17β in plasma after
smoothed curve fitting of mean values of at least 3 sheep per
treatment group. OE is oestradiol-17β (20 mg), P is progesterone (200 mg), T is testosterone (200 mg) and TBA is trenbolone acetate (140 mg).

It was concluded that both testosterone and progesterone influence the absorption of oestradiol-17β from combined implants in the same way as trenbolone acetate. Sustained absorption of oestradiol-17β was thought to be the reason for the greater growth rates in the second half of the treatment period reported in steers (Heitzman et al., 1981) and wethers (Harrison, 1981) treated with combined implants of trenbolone acetate and oestradiol-17β compared with those treated with the two steroids placed into ears as separate pellets. The commercial implants containing either oestradiol-17β or oestradiol-benzoate are given in Table 6.

TABLE 6. Implants of endogenous steroids.

Steroids	Dose	Trade name
Oestradiol-17β	ca 50 µg/d	Compudose
Oestradiol-17β Testosterone	20 mg 200 mg	Implixa-bf
Oestradiol-benzoate Testosterone propionate	20 mg 200 mg	Synovex-H
Oestradiol-17β Progesterone	20 mg 200 mg	Implixa-bm
Oestradiol-benzoate Progesterone	20 mg 200 mg	Synovex-S Steer-oid

The concentration of each steroid in plasma was higher in the jugular vein ipsilateral to the implant than in the contralateral vein. The difference between the concentrations in the two veins was used to calculate the biological half-lives of the steroids; for oestradiol-17β and trenbolone the mean values ranged from 1.8-6.8 min and 3-4 min respectively and for testosterone and progesterone the mean values were 4.8 and 3.5 min respectively (Harrison, 1981).

Residues

The residues in edible tissues of cattle of oestradiol, oestrone, testosterone and progesterone following implantation with the preparations in Table 6 have been reported

(Hendricks, 1981; Reid, 1981). A radioimmunoassay (RIA) method was used after lengthy extraction procedures. The methods were extremely sensitive and it was possible to measure concentrations of 5 ng/kg (5 parts per trillion or 5 in 10^{12}) of steroid.

The data from these studies was used to construct Table 7 to compare the intake of residues in meat with the daily production of endogenous steroids in the human. Under the conditions recommended for the use of implants of endogenous steroids, treatment of animals resulted in residues in meat that are orders of magnitude lower than those found occurring naturally in bulls and pregnant cows.

TABLE 7. Human intake of anabolic steroids from meat compared with their endogenous production in humans of various ages.

Production in humans (µg/d)	Testos-terone	Oestrogen	Progesterone
Adult male	6480	136	416
Women - range during cycle	240	190-1600	418-19600
late pregnant	320	64300	294000
post-menopausal	140	46	326
Pre-pubertal child	32	42	150
Maximum amounts of hormone (µg) in 250 g meat			
Untreated cattle	0.13[a]	0.11[b]	2.5[b]
Treated steer	0.0006	0.005	0.15
Treated heifer	0.025	0.005	-

[a]Mature bull
[b]Pregnant cow
Data from Hendricks (1981) and Reid (1981)

REFERENCES

Agthe, von O. 1980. Die Anwendung des Radioimmunoassays für diäthylstilböstrol auf Kotproben von Mastkälbern. Archiv. fur Lebensm. 31, 102-105.

Dixon, S.N. 1980. Radioimmunoassay of the anabolic agent
    zeranol. I. Preparation and properties of a specific
    antibody to zeranol. J. Wet. Pharmacol. Therap. 3,
    177-181.
Dixon, S.N. & Heitzman, R.J. 1981. Residues of anabolic
    agents in beef cattle and sheep. In "Anabolic Agents in
    Beef and Veal Production" EEC Workshop, ISBN-0-905442-
    54-7, pp. 58-69.
Harrison, L.P. 1981. In "An Investigation into the Effects of
    Anabolic Steroids on Ruminants". Ph.D. Thesis,
    University of Reading.
Harwood, D.J., Heitzman, R.J. & Jouquey, A. 1980. A
    radioimmunoassay method for the measurement of residues
    of the anabolic agent hexoestrol in tissues of cattle
    and sheep. J. Vet. Pharmacol. Therap. 3, 245-254.
Heitzman, R.J., Gibbons, D.N., Little, W. & Harrison, L.P.
    1981. Comparison of the performance of beef steers
    implanted with the anabolic steroids trenbolone acetate
    and oestradiol-17β alone or in combination. Anim. Prod.
    32, 219-222.
Heitzman, R.J. & Harwood, D.J. 1977. Residue levels of
    trenbolone and oestradiol-17β in plasma and tissues
    of steers implanted with anabolic steroid preparations.
    Br. Vet. J. 133, 564-571.
Heitzman, F.J. & Harwood, D.J. 1982. The radioimmunoassay
    (RIA) of hexoestrol residues in faeces, tissues and
    body fluids of bulls and steers. Vet. Rec. In Press.
Hendricks, D.M. 1981. Assay of naturally occurring oestrogens
    in bovine tissues. In "Steroids in Animal Production"
    (Ed. H. Jasiorowski) (Warsaw Agric. Univ. - SGGW-AR -
    Roussel-Uclaf, Warsaw), pp. 161-170.
Herriman, I.D., Harwood, D.J., Blandford, T. & Lindsay, D.
    1982. The distribution of hexoestrol residues in
    caponised chickens. Vet. Rec. 111, 435-436.
Hoffmann, B. 1978. Use of radioimmunoassay (RIA) for
    monitoring hormonal residues in edible animal products.
    J. Ass. Off. Anal. Chem. 61, 1263-1273.
Hoffmann, B. & Laschütza, W. 1980. Entwicklung eines radio-
    immunotests zur Bestimmung von Diäthylstilböstrol
    im Blutplasma und Essbaren Geweben vom Rind. Arch. für
    Lebensm. 31, 105-116.
Karg, H. & Vogt, K. 1981. Residues of diethylstilboestrol
    (DES) in veal calves. In "Anabolic Agents in beef and
    Veal Production", EEC Workshop ISBN 0-905442-54-7,
    pp. 70-83.
Phillips, M. & Harwood, D.J. 1982. An assay for 17α-OH trenbo-
    lone, the main biliary metabolite of trenbolone acetate
    in cow. J. Vet. Pharmacol. Therap. In press.
Pottier, J., Coustry, C., Heitzman, R.J. & Reynolds, I.P.
    1981. Differences in the biotransformation of a
    17β-hydroxylated-steroid, trenbolone acetate, in rat
    and cow. Xenobiotica. 11, 489-500.
Reid, J.F.S. 1981. Human health aspects of natural hormones
    used as growth promoters in cattle. In "Anabolizzanti
    in Zootecnia e Salute Pubblica". Soc. Ital Biuratria,
    pp. 142-155.

14

Rico, A. 1982. Metabolism of anabolic hormones in target species. European Toxicology Forum, Geneva. pp.327-338.

Riis, P.M. & Suresh, T.P. 1976. The effect of a synthetic steroid (trienbolone) on the rate of release and excretion of subcutaneously administered estradiol in calves. Steroids. 27, 5-12.

Ross, D. 1981. Toxicology and residues of trenbolone acetate as a model. In "Steroids in Animal Production" (Ed. H. Jasiorowski), Warsaw Agric. Univ., SGGW-AR, Roussel-Uclaf, Warsaw, pp. 227-234.

Stephany, R. 1982. The detection of diethylstilboestrol (DES) in the urine of veal calves and cattle via various chemical methods. A comparative study in Netherlands during 1981. Report from Rijks Institute of Public Health, Bilthoven, Holland.

Verbeke, R. 1979. Sensitive multi-residue method for detection of anabolics in urine and in tissues of slaughtered animals. J. Chrom. 177, 69-84.

Wagner, J.F. & Pankhurst, J.W. 1981. The use of hormones in animal production. Report from Eli Lilley Research.

DISCUSSION ON DR. HEITZMAN'S PAPER

Dr. Verbeke: The antibody is specific against zeranol but the main metabolite is zeralanone in ruminants. What use is this antibody for monitoring and control purposes?

Dr. Heitzman: The main metabolite is zeralanone and our first antibody cross reacts 100% with it. It occurs as about 10% of total metabolites. We do not have accurate data on the residues of the metabolites of zeranol. We could use the antibody to detect the major residues of zeranol in tissue. If we wished to characterise the metabolites of zeranol we could use the second antibody described. Using HPLC we could check for both metabolites if we wished. I don't think the issue of zeralanone is very important since it only accounts for 10% of the metabolites of zeranol.

Dr. Buttery: Is it the α or β metabolite of trenbolone that is active for growth promotion?

Dr. Heitzman: Trenbolone acetate is the 17β acetate and it is hydrolysed to 17β TBOH which is ten times more potent than the α metabolite in the mouse as an androgen.

Prof. Karg: Can the enzymes in ruminants only convert trendione to the 17α trenbolone?

Dr. Heitzman: There may be some conversion of 17β trenbolone to the ketone in blood.

Dr. Forbes: Had the sheep, cattle and goats you referred to been pretreated with trenbolone and the rats, monkeys, dogs not?

Dr. Heitzman: No.

<u>Prof. Hoffmann:</u> 17α epimerisation is typical for ruminants. They convert oestrone into 17α oestradiol and testosterone into 17α testosterone. (epitestosterone).

ASPECTS ON TOLERANCE LEVELS OF ANABOLIC AGENTS WITH
SEXHORMONE LIKE ACTIVITIES IN EDIBLE ANIMAL TISSUES

B. Hoffmann

Institut für Veterinärmedizin (Robert von Ostertag-Institut)
des Bundesgesundheitsamtes Berlin

INTRODUCTION

The use of compounds with sexhormone like activities as anabolic agents
in food animals has been an agricultural practice for more than 30 years
(see Hoffmann et al., 1975b). While some countries, for example the USA
and Great Britain, have given approval to legal treatments, other countries strictly prohibited the use of these compounds, in particular the
use of "oestrogens" (see Hoffmann, 1982). Thus a divergence of views towards the application of anabolic sex hormones is apparent and a public
discussion stired up on the human health aspects, particularly in those
countries exerting a strict ban. This was well demonstrated during the
recent EEC-oestrogen scandal in 1980/1981.

In simplifying the situation the two basic and apparently not answerable
questions were:
"Can residues of anabolic sex hormones at all be tolerated and if it
were so, what are the tolerance levels?"
This paper is an attempt to give an answer, based on our present
scientific knowledge and understanding.

GENERAL CONSIDERATIONS

Anabolic sexhormones exert distinct pharmacological and biological ac-
- tivities. From a scientific point of view there are no reasons, why
their evaluation in respect to efficacy, as well as animal health - and
public health aspects, should not follow the same guidelines as applied
for the regulation of veterinary drugs or feed additives, for example.

Such guidelines have been established officially or sometimes less officially in most countries. However, it has to be accepted that they need
continuous updating, particularly in respect of the scientific developments in pharmacology and toxicology and the understanding of these
achievements. Thus different criteria have been applied between the
countries at a given point of time or even within one country in the
course of time.

By accepting the Council Directives on Veterinary Medicinal Products
(81/ 851/EEC and 81/852/EEC) a major step forward in respect to harmoni-
zation within the EEC-member states has been made. It can be considered
a logical consequence, that an EEC scientific working group on anabolic
agents in animal production decided, that the most suitable criteria
for assessing the safety of anabolic agents (growth promoters) were
those set out in Part. 2, Chapter I, Section A-D of the Annex of Directive 81/852/ EEC of 28 September 1981 on the approximation of laws of
the Member States relating to analytical, pharmaco-toxicological and
clinical standards and protocols in respect of testing of veterinary
medicinal products.

SAFETY EVALUATION IN RESPECT TO PUBLIC HEALTH

The task to do such a safety evaluation should only be undertaken, if
adequate efficacy of the drug on the claimed field of application has
been demonstrated and if it seems to be acceptable in respect to animal
health. In general the benefit or efficacy of anabolic sexhormones has
been well demonstrated (for review of literature see Lu, Rendel, 1976,
Proc. EEC Workshop 1981[1], Jasiorowski, 1981) and the use of these
drugs also seems to concur with animal health aspects when used as suggested, that is in slaughter animals only and not in animals intended
for breeding purposes.

---

[1] Anabolic Agents in Beef and Veal Production. Commission of the
European Communities, Brusseles, ISBN-0-905442-54-7

In Table 1 is shown how anabolic sexhormones can be classified according to their hormonal activity and chemical moiety or origin.

Table 1: Classification of Available Anabolic Sexhormones

Structure/ origin	Hormonal Activity		
	Oestrogenic	Androgenic	Gestagenic
Endogenous steroid	+	+	+
Steroidal xenobiotic	+	+	+
Non-steroidal xenobiotic	+		

Basically the same guidelines for safety evaluation can be applied for all compounds classified in Table 1. Primary goal of such a safety evaluation is to assess the pharmacological and toxicological properties of residues in edible animal tissues at the time of slaughter following treatment of the animal. This evaluation also requires adequate information on the metabolism and pharmacokinetic profile of the drug. Figure 1 shows the three major possibilities which might be uncovered in such an investigation (Residue Study) and the resulting major consequences for the demands on "residue pharmacology and toxicology".
As it was established during the past 10-15 years as a result of residue analysis, a distinction can be made between those anabolic sexhormones which are identical with the endogenous steroids and the xenobiotic-type anabolics.

20

Fig. 1: Outline of possible results of residue study and
consequences for "residue pharmacology and toxicology"

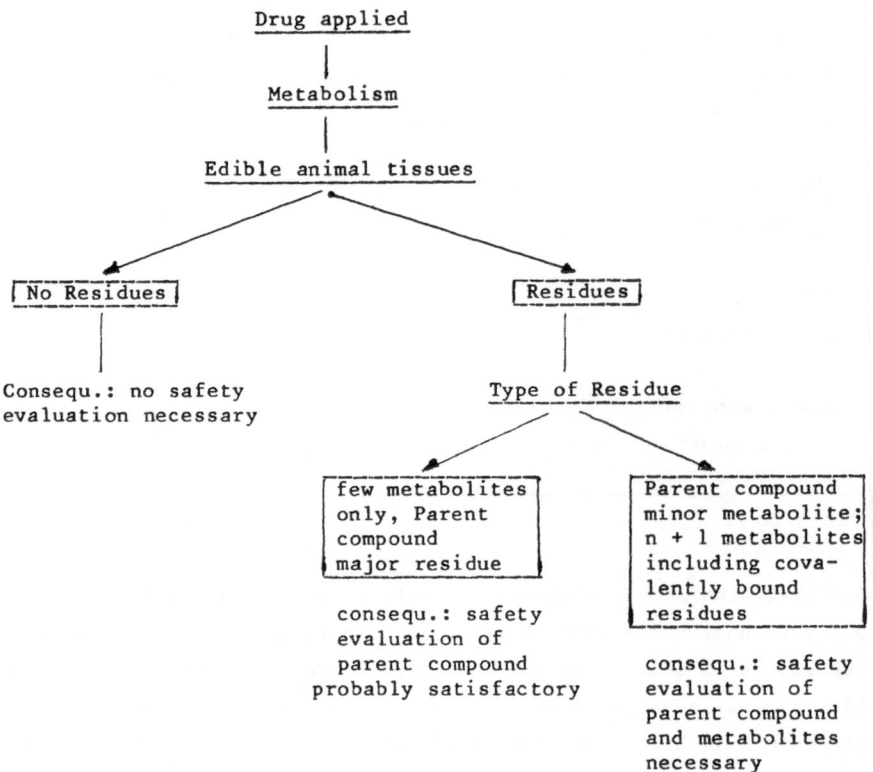

Endogenous Steroids:

The three endogenous steroids also used in anabolic preparations are oestradiol-17ß (estra-1,3,5(10)-triene-3,17ß-diol), testosterone (17ß-hydroxyandrost-4-en-3one) and progesterone (pregn-4-ene-3,20-dion). Also those esters readily yielding the free steroid when entering the circulation, like oestradiol benzoate or testosterone propionate, should be considered as endogenous steroids. They are the major sex hormones in man and animals exerting important physiological functions, lastly responsible for the survival of the species. They are present in all individuals. However, the production rate and hence the plasma levels vary depending on sex, age and physiological status (Velle, 1976, Hoffmann, 1981a, Vermeulen 1976/1981). Thus it has to be expected, that these steroids will also occur as natural constituents of food of animal origin in relation to the endogenous sex steroid production and the physico-chemical properties of these compounds. When used as anabolic implants in cattle (veal calf, heifer, steer) the total dose per animal given at one time usually does not exceed 20 mg for oestradiol-17ß and 200 mg for testosterone or progesterone. Following these treatments and observing the recommended withholding periods (in general 60-90 days) it was not possible to distinguish at slaughter between treated and untreated animals, based on the concentrations of these sexsteroids measurable in edible tissues (Hoffmann et al., 1975a, b; Hoffmann, Karg, 1976; Hoffmann, Rattenberger, 1977; Henricks, Torrence, 1978; Hoffmann, 1978; Reid, 1980; Henricks, 1981; Hoffmann, 1982; Meyer, 1983; Karg et al., 1983).

As is obvious from the data summarized in table 2, the steroid conentrations determined in tissues of treated animals are well at the lower part of the physiological range, based on the "inter-animal variation". Also within groups (calves, steers etc.) the treatment has only marginal effects on the tissue concentrations. At any rate the by far higest levels were observed in the sexually mature animals, particularly in the bull (testosterone) and the pregnant heifer or cow (oestrogens, progesterone).

These findings have clearly demonstrated, that a situation can come up
following the use of endogenous sex steroids, where as a consequence
of the physiological background levels, a residue problem simply does
not exist. However, whenever such a situation does occur, the question
of "residue pharmacology and toxicology" is not relevant.

All other aspects related to the safety of naturally occuring steroids
(of endogenous or exogenous origin) are secondary. Yet they should not
be disregarded because they in fact may be used to further define the
safety margin. Some important points in this respect are as follows
a) endogenous steroids have a low bioavailability following oral consumption because they are readily biograded when entering entero-hepa-
tic circulation in man and animals.
b) in relation to the endogenous steroid hormone production in the human
and even when considering the most sensitive segment of the population,
prepubertal children (Bidlingmeier, Knorr, 1978; Vermeulen, 1981), it
can not be expected that the comparatively minute amounts of steroid
hormones which are consumed more or less regularly with edible tissues,
might have a chance to measurably interact with the endocrine regulatory
mechanisms.
c) ever since mankind has developed, food of animal origin was a major
nutritive source. Apparently the anabolic endogenous steroids consumed
so far with these edible tissues have been tolerated quite well.

Anabolic Compounds of Xenobiotic Nature with Sex Hormone
like Activities:

General aspects: Not only in theory but also in practice the "no-residue"
concept can not be applied to these compounds. Thus for regulatory purposes it is necessary to evaluate the safety of residues and to see whether tolerance levels can be established. (This would also apply to the
situation if the use of endogenous steroids in general would give rise
to residue levels above the physiological range!).

Table 2: Average Concentrations (pg/g) of Testosterone, Progesterone and of Oestrogens in Tissues of Treated and Untreated Cattle (taken from: Hoffmann, Karg 1976; Hoffmann, Rattenberger 1977; Henricks 1981, Reid 1981 and Meyer 1983)

Compound determined	Animal	Tissue examined			
		Muscle	Liver	Kidney	Fat
Testosterone	Bull	535	749	2783	10950
	Heifer	92	193	595	250
	Veal calf	16	39	256	685
	" (treated)[a]	70	47	685	340
	Steer [e]	101			
	" (treated)				
Progesterone	Pregnant-cow				360200
	Heifer				16700
	Veal calf				5800
	" (treated)[b]				12500
	Steer[e]	124			
	" (treated)				
Oestradiol-17ß	Pregnant-cow	370-860			–
	Steer	14,4	12,0	12,6	–
	Heifer	12,0	38,3	39,8	–
Oestrone	Pregnant-cow	120-2090			
Total oestrogens	Veal calf	2	600	270	–
	" (treated)[d]	7	840	320	–
	Steer	12,5	27,6	26,0	–
	" (treated)[e]	21,8			
	Heifer	13,0	71,0	70,8	

[a] Slaughtered 77 days after implantation of 20 mg oestradiol-17ß + 200 mg testosterone.
[b] Slaughtered 70 days after implantation of 20 mg oestradiol-17ß + 200 mg progesterone.
[c] Slaughtered 70-77 days after implantation of 20 mg oestradiol-17ß + 140 mg TBA.
[d] Slaughtered 60 days after implantation of 20 mg oestradiol-benzoate + 200 mg testosteronepropionate.
[e] Slaughtered 66 days after implantation of 200 mg oestradiol-benzoate + 200 mg testosteronepropionate or 200 mg progesterone.

A common practice 'en route' to set tolerance levels is to check whether
an amount of residue can be defined which most likely can be consumed
daily without affecting the consumer's health. The formula to calculate
such an Acceptable Daily Intake (ADI) is a follows:

$$ADI = \frac{NEL \times bodyweight\ average\ consumer}{S\ F}$$

In this formula NEL stands for No-Effect Level and SF for Safety Factor.
The NEL can be defined as the oral dose of a drug which exerts no further pharmacolcogical or toxicological effects, for example in a sub-
chronic or chronic toxicity study or in another adequate investigation.
Depending on the metabolism of a drug in the target animal species, the
NEL might not only have to be established for the parent compound but
also for one or several metabolites (see fig. 1). In any case such a
study should also clearly outline the toxicological and pharmacological
profile of a compound- or of a residue. (Thus the so called relay toxicity studies where residue containing tissues are fed to laboratory animals can be considered of limited value only, since the doses reaching
the animal usually don't allow to establish such a profile). If a NEL
cannot be defined, the ADI-value and hence the tolerance level cannot
be set and the drug should not be used.

By convention the SF is usually set at 100 if no specific toxicological
considerations have to be taken into account. However, in case of a particular toxicity, for example teratogenicity, the SF is usually elevated
to 1000 or more. Again if for example due to a lack of information a SF
cannot be established, no ADI or tolerance levels can be set.

The ADI-value itself is based on an estimated daily consumption rate
(foodbasket) which may vary between countries depending on eating habits; to give an example it is quite common to calculate with an average
daily consumption of 300 g muscle, 100 g liver, 50 g kidney and 50 g
fat (Hoffmann, 1981b, Somogyi, 1982).

Special aspects: Many studies have shown that most sex hormones investigated exhibit a carcinogenic potential at varying dose levels in chronic toxicity studies (see: IARC Monographs on the Evaluation of carcinogenic Risk of Chemicals to Man, WHO - Lyon, 1974).

However, on the other side it is well established that sexhormones or compounds with sexhormone like activities have a basic promoting potential. Thus the question to be answered is, whether the tumorogenic or carcinogenic effects seen are due to the epigenetic (hormonal) or genotoxic potential of the compound (see also Summary and Conclusion; EEC-Workshop, 1981[2]; Ashby, 1982; Report on a WHO-Working Group, 1982[3]). To elucidate this situation adequate information on the mechanisms of action of a compound as well as on its mutagenic activity and other parameters have to be available. It is obvious that at present neither a NEL nor a SF can be delineated for a true genotoxic carcinogen. However, if the observed tumorogenic effect seems to be of epigenetic nature, a NEL might be calculated. In respect to sexhormones and compounds exhibiting sexhormone like activities, such a NEL might be identical with the No-Hormonal-Effect Level because the promoting activities relate to the pharmacological (hormonal) properties of the compounds.

For the compounds discussed the most at present, the situation can be briefly summarized as follows:

a) Stilbene oestrogens

The stilbene oestrogen best investigated is dietylstilboestrol (DES). A review on its toxicity has been published recently (Metzler, 1982). From there it can be taken that the genotoxic potential of DES has been established not only by carcinogenicity and mutagenicity studies but also by epidemiology. Thus no tolerance level can be set for residues of this compound in edible animal tissues.

----

[2] See footnote no 1
[3] WHO-Euro Reports and Studies 59 (1982),
    Health aspects of residues of anabolics in meat

Due to structural similarities and a lack of adequate information it presently has to be assumed, that all silbene oestrogens exhibit similar properties. The use of stilbene oestrogens as growth promoters in most countries of the world is at present strictly prohibited, which is more than just not tolerated.

b) Trenbolone (TBOH)

TBOH is a steroidal anabolic agent with an androgenic potential higher than testosterone (Neumanh, 1976). Basic information on its metabolism in the target animal species is available (Pottier et al., 1981; Rico et al., 1981; Schopper, Hoffmann, 1981). However, the occurrence of co-valently bound residues (Ryan, Hoffmann, 1978) still gives rise to some not yet fully answered questions. From the literature published so far (Ashby, 1982; Roe, 1981; Ross, 1981) and the personal information made available, the tumorogenic potential of TBOH seen in one study at a high dose level seems to be of epigenetic nature. However, no final conclusions in respect to a no-hormonal-effect level can yet be drawn and therefore the question of an ADI and tolerance level is still open. Thus according to latest standards final conclusion in respect to the use of this compound are still pending.

c) Zeranol

Zeranol is a derivative of a mycooestrogen with relatively weak oestrogenic activity. The drug has been available since 1969 and so far as is obvious from published (for review of literature see Brown, 1980/ 1981) and unpublished material, safety evaluation done previously in many points does not meet modern standards. Thus also in this case a lack of adequate information does not allow further conclusions at present.

SUMMARY AND CONCLUSIONS

Evaluation of the pharmacological and toxicological properties of residues of anabolic agents in respect to public health should be performed according to the latest criteria laid out for the evaluation of veterinary drugs. In principle no "a priori" differentation is justified between xenobiotic agents and naturally occuring substances. However, in the case of the endogenous steroids oestradiol-17ß, testosterone and

progesterone, convincing evidence has been presented, that following recommended treatments and based on the tissue-hormone levels measurable,
no differentiation can be made at the time of slaughter between treated
and untreated animals. Thus a residue problem does not exist and no further questions in respect to public health have to be answered. A further safety margin is obtained by the low oral bioavailability of these
steroids and the fact, that in respect to production of sex steroids
in the human himself, the amounts consumed with food of animal origin
seem to be negligible. In the case of xenobiotic anabolic agents safety
evaluation should lead to the establishment of an ADI-value and hence
of a tolerance level. A prerequisite for this is that a true no-effect-
level and an adequate safety factor can be defined, which - however -
at present is not possible in the case of compounds where carcinogenicity seems to be based on their genotoxic potential; the exclusion of
diethylstilboestrol from being used in food animals is based on this
observation. The situation is different for these compounds where a
more or less apparent tumorogenic potential can clearly be related to
their epigenetic potential; there it seems possible to delineate a no-
effect-level based on the pharmacological properties of the drug, which
for anabolic sex steroids would be their hormonal activity (no-hormonal-
effect-level). Evidence has been presented that this situation most likely will apply for some of the anabolics presently being used or under
investigation, like for example trenbolone; however, further information
has to be created for a final assessment according to recent criteria.

REFERENCES

Ashby, J. (1982): Genotoxicity and in vitro effects of hormones.
Proc. European Toxicology Forum, May 3-7, 1982, Geneva, Switzerland,
pp 339-346.

Bidlingmaier, F. and Knorr, D. (1978): Oestrogens, Physiological
and Clinical Aspects. Pediatric and Adolescent Endocrinology,
Vol. 4, S. Karger, Basel, München.

Brown, R.G. (1980): Toxicology and tissue residues of Zeranol. in
The Use, Residues and Toxicology of Growth Promoters, ISBN 0-
905442-44.x, pp 31-37.

Brown, R.G. (1981): Toxicology and residues of Zeranol.
Proc.: Anabolizzanti. In Zootecnia E Salute Pubblica, Societa
Italiana die Buiatria, 5.6.1981, 97-115.

Henricks, D.M. and Torrence, A.K. (1978): Endogenous estradiol-17ß
in bovine tissues. J. Assoc. Off. Anal. Chem. 61, 1263-1273.

Henricks, D.M. (1981): Assay of Naturally Occurring Oestrogens in
Bovine Tissues, in Steroids in Animal Production. H. Jasiorowski,
Ed., ISBN 83-00-01686-4, ARS POLANA-RUCH, P.O.Box 1001, 00-068
Warsawa, pp 161-170.

Hoffmann, B. (1978): Use of radioimmunoassay (RIA) for monitoring
hormonal residues in edible animal products. J. Ass. Off. Anal.
Chem. 61, 1263-1273.

Hoffmann, B. (1981a): Levels of endogenous anabolic sex hormones in
farm animals. In: Proc. Anabolic Agents in Beef and Veal Production, Commission of the European Communities, Brussels, ISBN 0-
905442-54-7, pp 96-112.

Hoffmann, B. (1981b): Pharmakologisch wirksame Stoffe für Nutztiere - Östrogenproblematik. Bayer. Landw. Jahrbuch 58, SH 1, 5-57.

Hoffmann, B. (1982): The use of hormonesinfood-producing animals. In: Health aspects of residues of anabolics in meat, WHO-EURO Reports and Studies 59, pp 16-27.

Hoffmann, B. and Karg, H. (1976): Metabolic fate of anabolic agents in treated animals and residue level in their meat. In: Environmental quality and safety. F. Coulston, F. Corte, Eds., Suppl. Vol 5, Anabolic agents in animal Publ. production, Georg Thieme, Stuttgart, Germany, p.p. 181-191.

Hoffmann, B. and Rattenberger E. (1977): Testosterone concentrations in tissue from veal calves, bulls and heifers and in milk samples. J. Anim. Sci. 46, 635-641

Hoffmann, B., Karg, H., Heinritzi, K.H., Behr, H. und Rattenberger, E. (1975a): Moderne Verfahren der Oestrogenbestimmung und deren Anwendung für die Rückstandsproblematik. Mitt. Gebiete Lebensm. Hyg. 66, 20-37.

Hoffmann, B., Karg, H., Vogt, K., Kyrein, H.J. (1975b): Aspekte zur Anwendung, Rückstandsbildung und Analytik von Sexualhormonen bei Masttieren. Forschungsbericht der DFG; Rückstände in Fleisch und Fleischerzeugnissen, Harald Boldt-Verlag KG, Boppard, pp 32-59.

Jasiorowski, H (1981): Steroids in Animal Production, ARS POLANA-RUCH, P.O.Box 1001, 00-)68 Warzawa, Poland.

Karg, H., Meyer, H.D., Vogt, K., Hoffmann, B., Landwehr, M. and Schopper, D. (1983): Residues and Clearance of Anabolic Agents in Veal Calves. These Proceedings.

30

Lu, F.C. and Rendel, J. (1976): Anabolic agents in animal production. Environmental quality and safety. Suppl. Vol. 5, Georg Thieme Verlag, Stuttgart, Germany.

Metzler, M. (1981): Residues of Anabolics in Meat: Risks for Consumers. WHO Euro Reports and Studies 59, Health aspects of residues of anabolics in meat, pp 28-35.

Meyer, H.D. (1983): Neue Methode zur Bestimmung von Steroidöstrogenen in Geweben. Wiener Tierärztl. Wschr., in press.

Neumann, F. (1976): Pharmacological and endocrinological studies on anabolic agents. In: Environmental Quality and Safety, F. Coulston, F. Corte, Eds., Suppl. Vol 5., Anabolic Agents in Animal Production. Georg Thieme, Stuttgart, Germany, pp 253-264.

Pottier, J., Cousty, C., Heitzman, R.J. and Reynolds, I.P. (1981): Differences in the biotransformation of a 17ß-hydroxylated steroid, trenbolone acetate, in rat and cow. Xenobiotica 11, 489-500.

Reid, J.F.S. (1980): Significance of natural oestrogen-implanted beef to human health. The Use, Residues and Toxicology of Growth Promoters, ISBN-0-905442-44-x, pp 24-30.

Rico, A.G., Burgat-Sacaze, V., Braun, J.P. and Bernard, P. (1981): Metabolism of endogenous and exogenous anabolic agents in cattle. In: Anabolic Agents in Beef and Veal Production, Commission of the European Communities, Brussels, ISBN-0-905-442-54-7, pp 45-56.

Roe, F.J.C. (1981): Toxicological aspects of anabolic hormone use in meat production. Proc.: Anabolizzantie In Zootecnia E Salute Pubblica, Societa Italiana die Buiatria, 5.6.1981, 159-169.

Ross, D.B. (1981): Toxicology and tissue residues of an exogenous androgen. Proc.: Anabolizzantie In Zootecnia E Salute Pubblica, Societa Italiana die Buiatria, 5.6.1981, 117-127.

Ryan, J.J., Hoffmann, B. (1978): Trenbolone Acetate: Experiences with Bound Residues in Cattle Tissues. J. Assoc. Off. Anal. Chem. 61, 1274-1279.

Schopper, D. und Hoffmann, B. (1981): Identifizierung von 17a-Trenbo-lon als Hauptausscheidungsprodukt des Trenbolonacetat-Stoffwechsels beim Kalb und sich daraus ergebende Konsequenzen für die Rückstands-analytik. Arch. f. Lebensmittelhyg. 5, 141-144.

Somogyi, A. (1982): Gesundheitliche Bedeutung von Rückständen pharma-kologisch wirksamer Substanzen in der Tierhaltung. Bundesgesund-heitsblatt 25, 365-370.

Velle, W. (1976): Endogenous anabolic agents in farm animals. In: Environmental Quality and Safety, F. Coulston, F. Corte, Eds., Suppl. Vol. 5. Anabolic Agents in Animal Production, Georg Thieme, Stuttgart, Germany, pp 159-170.

Vermeulen, A. (1976): Plasma levels and secretion rate of steroids with anabolic activity in man. In: Environmental Quality Safety. F. Coulsten, F. Corte, Eds., Suppl. Vol. 5. Anabolic agents in animal production. Georg Thieme, Stuttgart, Germany, pp 171-180.

Vermeulen, A. (1981): Endogenous hormone levels in humans. In: Anabolic Agents in Beef and Veal Production. Commission of the European Communities, Brussels, ISBN-0-905442-54-7, pp 84-95.

DISCUSSION ON PROFESSOR HOFFMANN'S PAPER

Prof. Karg: We should emphasis one point. The positive evaluation of the physiological compounds is due to residue studies in edible tissues. When anabolic agents are properly administered in the ear, the implantation site does not enter the human food chain because the ear is discarded. In the future if anabolic agents are used in pigs what is the proper implantation site?

Prof. Hoffmann: Proper use in this case meant implantation in the ear, using the correct dose and observing the correct withholding period. The withholding period is zero as pointed out by Dr. Heitzman because the levels seen in these animals never exceed physiological levels e.g. the silastic rubber implant containing oestradiol.

Dr. Roche: From a consumer viewpoint, DES has been available as an anabolic agent for 25 years before being banned due to genotoxicity effects. Is it possible that current xenobiotic compounds on the market in the future could be classified similarly as new information from advanced technology becomes available?

Prof. Hoffmann: Yes, this comes up every week with various drugs used in veterinary medicine - e.g. nitrofuranes. There are also other drugs where we cannot come up with a no effect level, e.g. chloramphenicol. Thus, this is a common problem with many compounds. The question is what is the efficacy and need for such compounds. It is easier to forget about DES than chloramphenicol.

Dr. Heitzman: In the case of the xenobiotics the information available does not allow for setting of an ADI. As this information becomes available, through what route should it be channelled - e.g. JECFA or through the Commission or through national agencies?

<u>Prof. Hoffmann</u>: There are various national and
international routes. Within the EEC, however, it may be
be better to go through EEC channels because this has
been made an EEC issue already.

RESIDUES AND CLEARANCE OF ANABOLIC
AGENTS IN VEAL CALVES

H. Karg, H.H.D.Meyer, K. Vogt, M. Landwehr, B. Hoffmann[1] and
D. Schopper[2]

Institut für Physiologie der Südd. Versuchs- und Forschungs-
anstalt für Milchwirtschaft, Technische Universität München,
8050 Freising-Weihenstephan, Federal Republic of Germany

ABSTRACT
    The obvious benefit from the use of anabolic agents for
fattening veal calves has forced us to provide residue data
for risk evaluation and to establish analytical methods to
control illegal treatments. The requirement of extraction and
purification procedures being applied in connection with
radioimmunological and/or thin-layer-chromatographic determinations are presented. In model experiments, with the dis-
qualified compound diethylstilbestrol (DES), it could be
demonstrated that residues and clearance rate depend mainly on
the route of administration, the formulation of the preparation, the type of vehicle and the time after application.
After intramuscular injection of DES dissolved in oil a quick
clearance within one to four weeks has been observed, whereas
the application of a crystalline DES-suspension still gave
measurable values after three to four months. Due to easy
sampling and the relatively high DES concentration, faeces
should be used preferentially for monitoring purposes. Properly
used and dosed trenbolone acetate pellets (as licensed in
some countries already) gave TBOH concentrations -so far
extractable- below 1 nanogram/g in edible tissues and in the
nanogram range in excreta during a normal fattening
period. Exogenously administered physiological steroidal
oestrogens showed different clearance rates depending on the
site of implantation. The oestrogen levels in the muscle of
untreated as well as treated calves were in the low picogram
range while those in liver and kidney were in the higher picogram
range. Levels of these steroids of untreated and treated male
and female calves overlapped in some cases. The concentrations
of physiological steroids found in commonly consumed edible
tissues from adult cattle were very often higher than those of
properly treated veal calves.

---

present addresses:

[1] Institut für Veterinärmedizin des Bundesgesundheitsamtes,
   1000 Berlin 65, Federal Republic of Germany

[2] Institut für Tierhaltung und Tierzüchtung -470- der
   Universität Hohenheim,
   7000 Stuttgart 70, Federal Republic of Germany

INTRODUCTION

In some countries, especially on the European continent, there still exists a remarkable veal market. This type of meat production seems, irrespective of any economic forecasts, to remain unchallenged so long as consumers expect that restaurants should offer courses like "rôti de veaux", "vitello a la casa" or "Kalbsschnitzel". Producers, at least since about the past 1 1/2 decades, have been aware of the beneficial effect of anabolic agents in veal production. This is possible due to the lack of endogenous sexual hormones during the juvenile or prepuberal status of these animals. A discussion about the benefit / risk - evaluation in connection with the use of anabolic agents in general was promoted in recent years by the public. This concern occurred concomitantly with the detection of illegally treated veal calves and the occurrence of diethylstilbestrol (DES) residues in canned food containing veal.

The aim of this paper is to summarize the present status of residue data in edible tissues and excreta in order to allow the evaluation of the risk (given in the paper of Hoffmann within this program) and to consider reasonable monitoring measures.

We have to face the fact that without an efficient control system the illegal treatment of calves can not be excluded. Therefore, this paper will not only consider different compounds and formulations, but will also deal with practised routes of administration.

Data will be provided from the results of our investigations concerning 3 groups of sex-hormonal compounds: 1. Stilbene derivatives, 2. trenbolone and 3. natural steroids. Since according to our knowledge no relevant data are available on zeranol in veal calves this compound has been neglected in this paper.

METHODS

The majority of the results is based on radioimmuno-assay (RIA) data. The required sensitivity and the up-to-date availability of such systems for the determination of anabolic

TABLE 1: Reviewed methods for residue determination.

	required detection limit ng/g or ng/ml	required purification			final determination	
		extraction	phase-partition	chromato-graphy	RIA	TLC
P L A S M A						
DES	0.02	+	−	−	+	−
TBOH	0.02	+	−	−	+	−
Oestradiol	0.01	+	−	−	+	−
Testosterone	0.02	+	−	−	+	−
Progesterone	−	+	−	−	+	−
U R I N E						
DES	1.0	+	+	−	+	+
TBOH	1.0	+	−	−	+	+
Oestradiol	1.0	+	−	−	+	−
F A E C E S						
DES	5.0	+	+	−	+	+
TBOH	1.0-5.0	+	+	?	+	+
Oestradiol	1.0	+	+	+	+	−
L I V E R						
DES	0.1	+	+	−	+	−
TBOH	0.05-0.1	+	+	?	+	−
Oestradiol	0.1	+	+	+	+	−
Testosterone	0.1	+	+	+	+	−
K I D N E Y						
DES	0.1	+	+	−	+	−
TBOH	0.05-0.1	+	+	?	+	−
Oestradiol	0.1	+	+	+	+	−
Testosterone	0.1	+	+	+	+	−
M U S C L E						
DES	0.02	+	+	−	+	−
TBOH	0.05	+	+	?	+	−
Oestradiol	0.001	+	+	+	+	−
Testosterone	0.04	+	+	+	+	−
F A T						
TBOH	0.05	+	+	+	+	−
Oestradiol	0.02	+	+	+	+	−
Testosterone	0.04	+	+	+	+	−
Progesterone	−	+	+	+	+	−

TBOH: Oehrle, Vogt and Hoffmann, 1975; Hoffmann and Oettel, 1976;
Vogt, 1977; Schopper, 1981. Oestradiol: Karg, Hoffmann, Vogt and
Behr, 1972; Meyer, 1983. Testosterone: Hoffmann and Rattenberger, 1977.
Progesterone: Hoffmann, 1978.

sexual hormones have been recently summarized by Hoffmann and Blietz (in press). In addition we present a list of requirements concerning extraction and purification procedures (Table 1). Thin-layer-chromatographic procedures (Vogt, 1978, 1979) are also useful for residue determination in excreta (faeces and urine). Earlier studies carried out with a biological assay (mouse-uterus-weight-test) are still of confirmative interest if the chemical nature of the oestrogen is not required.

RESULTS AND DISCUSSION

1. DIETHYLSTILBESTROL (DES)

The data presented are based on RIA and extraction procedures described by Hoffmann and Laschütza, 1980, Vogt, 1980, Agthe, 1980, Vogt and Karg, 1982, and Hoffmann and Blietz (in press). Residues of DES found in edible tissues after different treatments of calves are summarized in Table 2.

The advantage of the liver and kidney versus muscle as marker tissue is well established. In comparison the quick elimination of DES from the orally treated calves (A and B), depending on the withdrawal time, was obvious. The extremely high concentrations of DES stored at the injection sites (D and E) demonstrate a possible source for substantial contamination of canned food by meat, if whole carcasses are processed. Experiment C demontrated that after intramuscular (i.m.) application of DES dissolved in oil the concentration of DES at the injection site was several times lower than after the administration of DES-dipropionate in a crystalline suspension (D and E) and also substantially lower than in former studies, where DES-dipropionate dissolved in oil was used. We reported previously (Vogt et al., 1970), that three weeks after i.m. injection of 75 mg DES-dipropionate, residues of bioactive DES were found to be 2.1 µg, 0.2 µg or 8 ng respectively at distances up to 5, 5-15 or 15-20 cm from the injection site. Recently, we (Vogt and Karg, 1982) found

TABLE 2: Residues of DES (free, conjugated or total) in edible tissues in veal calves after different types of application (results from 1 animal in all columns)

DES-concentrations (ng/g)

Tissue examined	A free	A conj.	B free	B conj.	C free	C conj.	D free	D conj.	E free	E conj.
Muscle	0	-	0	0.19	0.07	0.02	0.19	0	0.007	0
Liver	0.160	0.300	6.3	13.8	0.23	0.52	2.0	2.9	0.14	0.15
Kidney	0	-	3.6	6.9	0.08	0.54	1.0	2.2	0.10	0.27
					total		total		total	
Muscle at the injection site	-		-		17		143434		10290	

A) Short term oral; 5 mg DES/day; last supply 68 hours before slaughter

B) Short term oral; 5 mg DES/day; last supply 6 hours before slaughter

C) 200 mg DES i.m.; oily solution; slaughter 7 days after application

D) 150 mg DES-dipropionate in crystalline suspension i.m.; slaughter 23 days after application

E) 150 mg DES-dipropionate in crystalline suspension i.m.; slaughter 84 days after application

that samples taken only 1 cm apart may have differences in
DES-concentration of up to 3 or 4 decimals (Table 3) near the
injection site muscles (D and E) approximately 3 weeks or
3 months after i.m. administration of DES-dipropionate in
crystalline suspension. Therefore the chance of obtaining
striking results in samples taken from carcasses may be very
small. From the consumers point of view, perhaps " only "
one individual might be in danger consuming such high
amounts of DES in a piece of meat. However, in the case of
processed meat products, the injection site material will be
diluted and many more consumers might be endangered by
active amounts of DES. This problem indicates the requirement
for qualified sensitive methods for the detection of this
xenobiotic compound also in muscle far from injection sites.
At the same time it is also recommended to put emphasis
on more practical procedures for the determination of DES,
which can be used if excreta are available. Thus, in the
Federal Republic of Germany control measures are
preferentially directed to the living animals in the stables
of the producers. Excretion patterns for DES obtained in
different investigations are shown in Figures 1 - 5.
We prefer faeces instead of urine due to the fact that
DES-concentrations are about 1 decimal higher and that
it is easier to collect it from the living animal.

TABLE 3: DES-concentrations in muscle near the injection site
at different time intervals between treatment and
slaughter

Distance from injection site	Calf D (23 days)	Calf E (84 days)
	DES-concentration in muscle	
cm	ng/g	ng/g
- 6	-	1,43
- 5	-	0,40
- 4	-	0,27
- 3	-	0,72
- 2	6237	0,23
- 1	97047	880
0	143434	10290
+ 1	40309	5230
+ 2	13128	1850
+ 3	10463	75
+ 4	98	203
+ 5	27	1,50
+ 6	2	0,67
+ 7	-	1,40
Total residue	5,2 mg	0,37 mg

In Figure 1 the quick depletion (corresponding to data of
calves A and B, Table 2) after withdrawal of the orally fed
DES is shown.

Figures 2 and 3 demonstrate that the clearance of DES is
usually completed within 3 or 4 weeks if an oil vehicle is
used. This is in good agreement with previous studies by
Huis in't Veld et al. (1969) and recent investigations by
Boursier et al. (1982) and Hoffmann (1981). The latter found
a clearance of DES in faeces after i.m. application of 200 mg
free DES within almost one week. A similar response has been
shown for dienestrol (Hoffmann, unpublished data). However,
a tremendous delay in the excretion of DES has been observed
after the administration of DES-dipropionate in crystalline
suspension (Vogt and Karg, 1982 and unpublished data). There
was a low level of DES still measurable up to 4 months after
application. A higher initial dose only leads to a higher
clearance rate during the first weeks after treatment (Fig.4/5).

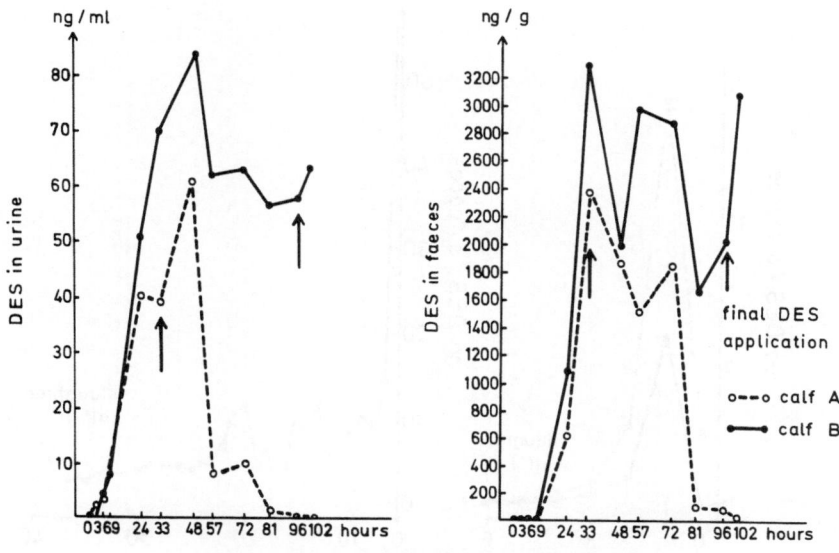

FIGURE 1: Excretion of DES in urine and faeces of 2 calves
after oral application of 2 x 2.5 mg DES/animal
and day.

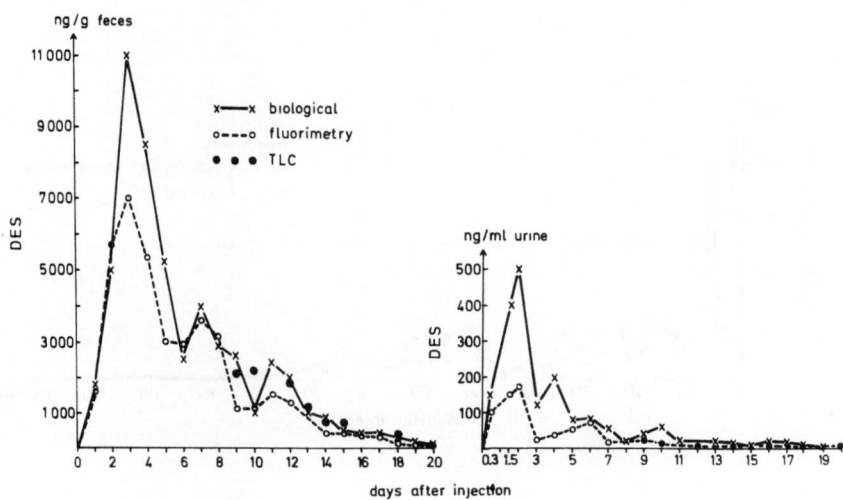

FIGURE 2: Excretion of DES in faeces and urine of 1 calf
after intramuscular injection of an oily solution
of 75 mg DES-dipropionate.

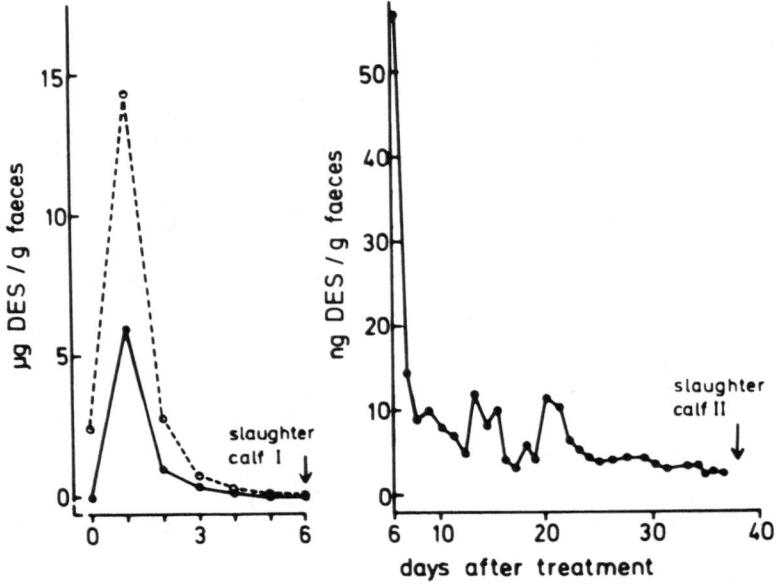

FIGURE 3: Elimination of DES with faeces in 2 calves following treatment with 200 mg DES, given i.m. in oily solution. o----o calf 1; ●——● calf 2.

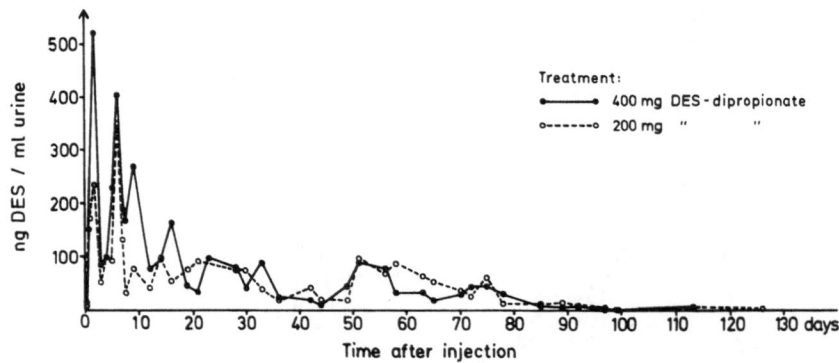

FIGURE 4: Excretion of DES in urine of calves after i.m. injection of DES-dipropionate in crystalline suspension.

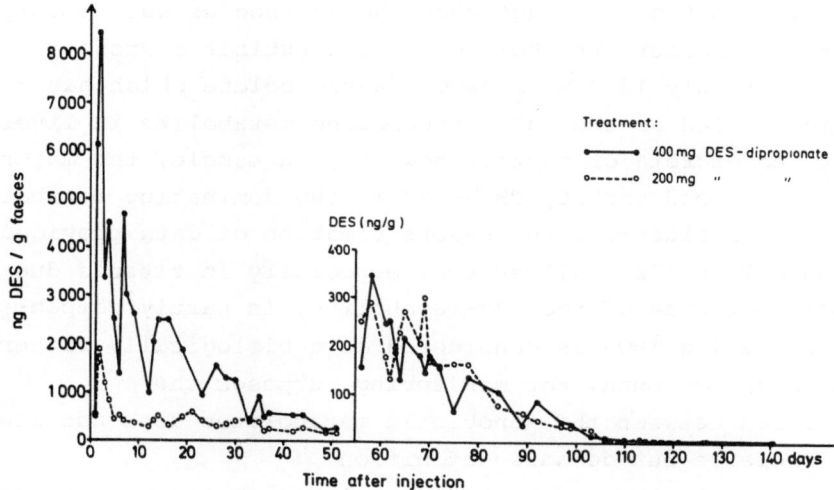

FIGURE 5: Excretion of DES in faeces of calves after i.m.
injection of DES-dipropionate in crystalline
suspension.

The results obtained with these two calves are in good agreement with those from a former experiment with 2 calves
receiving 150 mg of DES-dipropionate (Vogt and Karg, 1982).
These findings demonstrate that the galenic formulation is of
much more influence on the clearance pattern than the applied
dose. In forensic trials it is often important to reconstruct
from analytical data the possible time of the last treatment.
Thus it is recommended, as far as living animals are under
control, to perform DES-determination in two faeces samples
taken two or more weeks apart. A clearance curve which is
characterized mainly by the formulation of the administered
compound could then be approximately achieved.

2. TRENBOLONE

    In our evaluation of the data available from trenbolone
residue studies we have two major reservations:
1) The RIA systems as developed by Hoffmann and Oettel, 1976,

Schopper and Hoffmann, 1981 and Schopper, 1981 have to be
viewed in light of the fact that the antibodies were mainly
directed to measure trenbolone-17ß and exhibit a cross-
reaction of only 15-20% against 17a-trenbolone which has
been identified as the major trenbolone metabolite in liver,
kidney and excreta of cattle. However, in muscle, the major
part of the food basket, TBOH-17ß is the dominating metabolite.
For safety evaluations the underestimation of data provided
by a pure TBOH-17ß - RIA system, especially in tissues due to
the main presence of the 17a-metabolite, is partly compen-
sated since 17a-TBOH is considered as a biologically rather
inactivated compound. For monitoring purposes the
distinction between the xenobiotic metabolites does not seem
to be decisive but demands definition.

2) According to the information available at the moment any
analytical method based on extraction of tissues with organic
solvents seems to be insufficient to measure the apparently
covalently bound non-extractable residue. The non-extractable
residue exceeds by far the extractable residues in the case of
trenbolone (Ryan and Hoffmann, 1978, Schopper, 1981). With
these reservations we present ranges of residue data from
former experiments of total trenbolone (free plus conjugated)
in Table 4 (Hoffmann et al., 1976).

TABLE 4: Extractable TBOH-concentrations (min. and max. or
single values, respectively, in ng/g) in tissues of
female calves - 70 days after treatment with
140 mg TBA (I and II) and 20 mg oestradiol (only II).

Tissue	Group	
	I (n=3)	II (n=2)
Muscle	0.04 - 0.09	0.09 / 0.12
Liver	0.26 - 0.57	0.38 / 0.91
Kidney	0.08 - 0.19	0.28 / 0.43
Fat	0.17 - 0.25	0.48 / 0.46

In this experiment the normal dosages resulted in significant
detection of this xenobiotic compound 70 days after proper
administration. In an other experiment with two calves, also
treated with the normal dose of 140 mg trenbolone-acetate
each (at the ear base), slightly higher values were found
61 days after administration, but in the same order of
magnitude (Table 5).

TABLE 5: Extractable TBOH-residues of 2 calves after
        implantation of 140 mg TBA

Biological material		days after implantation		
		13	27	61 (slaughter)
Plasma	(ng/ml)	0.5/ 0.6	0.5/ 0.6	0.1/ 0.3
Urine	(ng/mg creatinine)	11.6/11.0	6.4/11.3	6.2/ 5.3
Faeces	(ng/g fresh faeces)	56.8/69.3	17.2/40.4	40.0/11.8
Muscle	(ng/g)			0.3/ 0.3
Liver	(ng/g)			1.5/ 1.6
Kidney	(ng/g)			0.4/ 0.4
Fat	(ng/g)			0.6/ 0.7

This Table also includes some results of the TBOH
determination in plasma, urine and faeces during the courses
of the investigations. These data indicate that urine and
faeces should be used preferentially for monitoring TBOH.
There were high values present within the range of measurement during the whole normal fattening period. Figures 6 - 8
give more related information. The evaluation of plasma
TBOH concentrations (Fig. 6), as demonstrated in an experiment
carried out in the Netherlands (Schopper et al., in
preparation), showed the striking effect of implant removal.
There was a quick drop of the contamination of the excreta to
the limit of detection (approximately 30 pg/ml). This level
was reached in the animal with the implants left in place only
after 8 - 9 weeks.

46

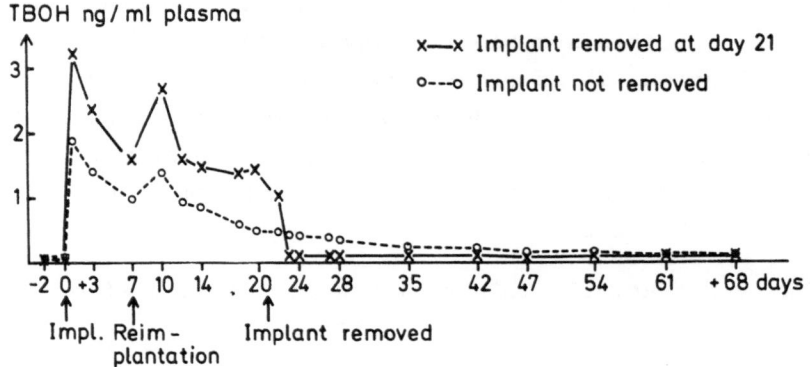

FIGURE 6: TBOH-concentration in plasma of calves with and
without removal of the TBA-implant.

Figure 7 (faeces) corresponds with data presented in Table 5
and shows the higher variability of the residue data during
the course of the experiment. More adjusted profiles were
obtained by residue determination of urine related to
creatinine (Figure 8).

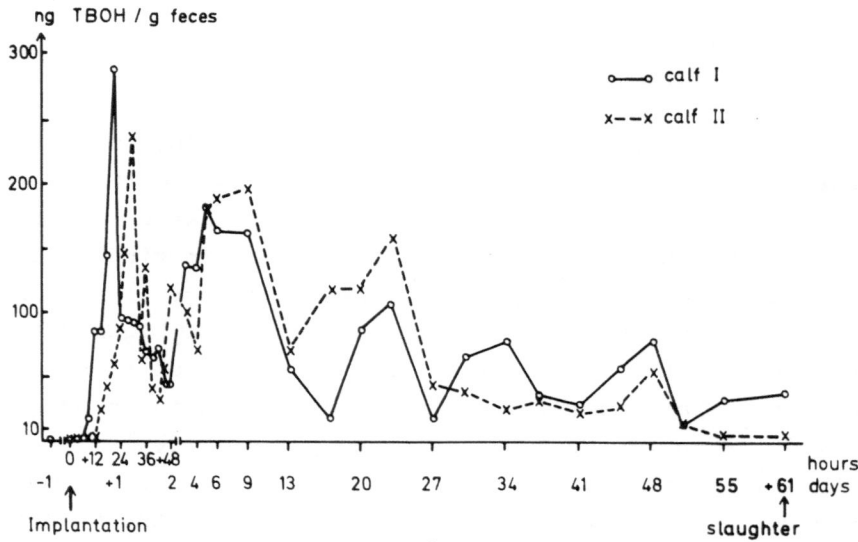

FIGURE 7: Excretion of TBOH in faeces of 2 calves implanted
with 140 mg TBA.

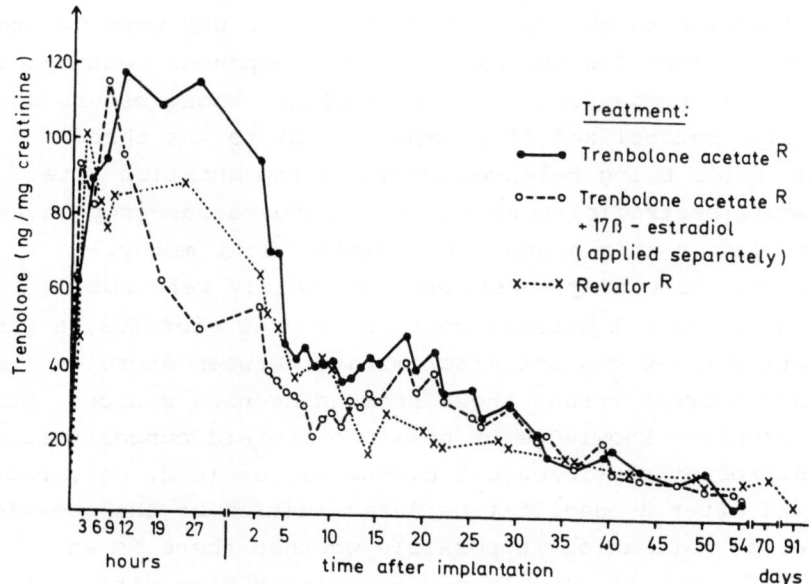

FIGURE 8: Excretion of trenbolone in urine of male
veal calves after subcutaneous implantation
of Revalor ®, TBA and 17ß-oestradiol
(administered separately) or TBA.

It should be noted that the latter Figure represents data from
thin-layer-chromatographic detections (Vogt, 1977). This is an
alternative procedure (also without distinction between
trenbolone-17a and -17ß), which is useful for monitoring
purposes if excreta are available. It is desirable to
establish for defined monitoring programs more exact quantitative relationships of the time courses of clearance after
administration of trenbolone-acetate, perhaps with a more
specific system for trenbolone-17a.

3. ENDOGENOUS SEX STEROIDS

According to the understanding of the WHO working group
(1982) endogenous sex steroids are the compounds oestradiol-17ß,
testosterone, progesterone and in addition those esters which
are rapidly metabolized (i.e. hydrolysed) to the three
steroids after being released from the implantation site.
For example oestradiol-benzoate and testosterone-propionate
are included in this group but compounds like methyl-
testosterone or ethinyl-oestradiol certainly were not
included. Since all tissues contain natural steroids, a simple
qualitative assay can not discriminate between steroids from
exogenous sources versus those from endogenous sources. Only
with a complete knowledge of possible steroid concentration
under different physiological circumstances (sex, age, race
etc.) and after a quantitative determination of the questioned
hormone, an estimation is possible whether there is an
exogenous source or not. In our experiment with different
preparations, doses and injection sites we tried to confirm
that suitable conditions for growth promotion with natural
steroids exist. On the other hand, monitoring systems are
needed to elucidate unsuitable conditions, i.e. illegal
treatments. Oestradiol-17ß and its physiological metabolites
stand in the focus of investigations of the mentioned 3
compounds due to the fact that they have (compared to some
synthetic oestrogens) greatly reduced but not negligible
oral activity. The bioavailability is comparatively
minimized for testosterone and perhaps negligible for
progesterone. Both compounds are generally combined in anabolic
preparations with oestradiol-17ß. Residue studies in muscle
(apart from the injection site), liver, kidney and fat of
veal calves are summarized in Table 6 for testosterone
and in Table 7 for total oestrogens. Progesterone values from
residue studies are available in fatty tissue, with mean
values of 6, 17 and 336 ng/g from untreated veal calves,
heifers or pregnant cows respectively and of 12 ng/g from
veal calves treated with 20 mg oestradiol-17ß + 200 mg
progesterone 70 days before slaughter (Hoffmann, 1978).

TABLE 6: Concentrations (pg/g) of unconjugated testosterone
in tissues of treated and untreated cattle.

	female calves control	calves treated[x]	heifers untreated	bulls untreated
	n = 5	n = 5	n = 3-5	n = 5
Muscle	11- 30	25- 127	59- 122	172-1663
Liver	15- 54	35- 63	111- 357	268-1233
Kidney	134-391	154-1476	178-1739	1000-5230
Fat	31-305	154- 774	198- 360	2444-21780

[x] Slaughtered 77 days after implantation of
20 mg oestradiol-17ß + 200 mg testosterone.

The results of Table 6 show high variation of testosterone
(Hoffmann and Rattenberger, 1977) in the picogram range.
Usually individual results from the treated calves overlapped
with those of the untreated animals. But levels in both groups
were substantially lower especially compared to those in normal
edible tissue from bulls.
In Table 7 we see that after all treatments, residues were in
the same order of magnitude. We have to keep in mind that
total oestrogens were determined which include oestradiol-17a
in addition to oestradiol-17ß and oestrone after hydrolysis
(i.e. in free and conjugated form).
Henricks (1977 and 1981) found lower oestradiol and "oestrogen"
concentrations in the liver and kidney of steers and heifers.
The differences to our results are not fully understood.
Besides the fact that the animals under investigation were not
directly comparable with our veal calves, we have to consider
that the major part (Dunn et al., 1977) of total oestrogens
in these tissues is represented by conjugated oestradiol-17a.

TABLE 7: Total steroidal oestrogens in edible tissues of veal calves (min.-max. or single values in pg/g). Treatment with oestradiol-benzoate (EB) either in pellets with testosterone-propionate or in oily solution.

| | Controls | | treatment 10 mg EB pellet s.c. | | | treatment 20 mg EB pellet s.c. | | treatment 50 mg EB in oily solution i.m. |
| | | | A | B | B | A | B | C |
	female n=7	male n=1	female n=6	female n=1	male n=1	female n=6	male n=1	female (freemartin) n=1
Muscle	<12- 20	-	<12	20	-	<12- 28	15	7 x)
Liver	120-1550	190	600-2150	465	50	580-1900	330	960
Kidney	120- 730	250	280- 410	1100	260	340-1520	560	940
Fat	40- 110	200	40- 90	140	90	90- 260	125	70

A) Site of administration on base of the ear; slaughter 75 days after administration.

B) Site of administration on ear cartilage; slaughter 62 days after administration.

C) Slaughter 3 weeks after intramuscular injection.

x) far from injection site, see Table 8.

In fact, as shown in Figure 9 (Schopper et al., in preparation), measurement in peripheral plasma of male calves
after treatment with oestradiol-17ß revealed elevations of
free oestradiol-17ß and (see higher range of the ordinate!)
especially of conjugated oestradiol-17a, but did not involve so
much oestrone, free oestradiol-17a or conjugated oestradiol-
-17ß. An alternative is the measurement of total oestrogens.
The example in Figure 9 represents evaluation with a RIA -
system including an antibody with a reduced crossreactivity
to oestradiol-17a and oestrone. Even if the profile for total
oestrogens gave reasonable average values and limited standard
deviations, large individual differences have to be considered.

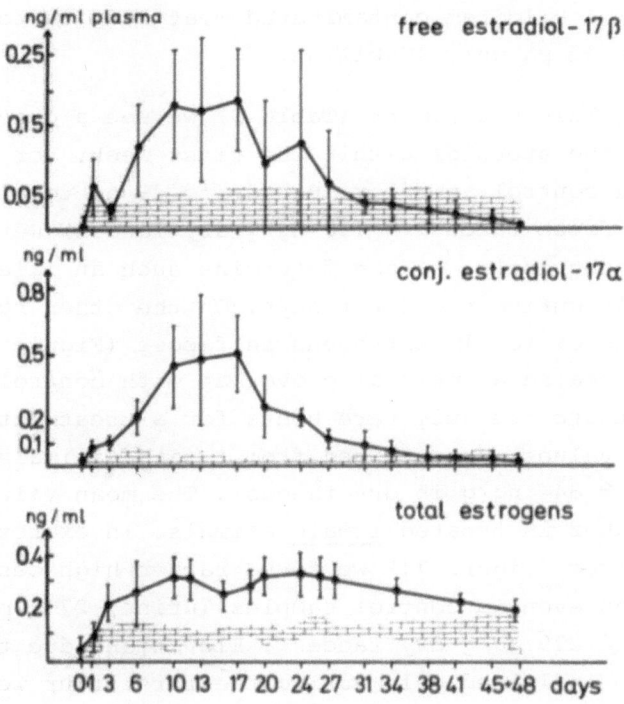

FIGURE 9: Oestrogens in plasma of male veal calves (n=6) after
implantation of 20 mg oestradiol-17ß in combination
with trenbolone-acetate s.c. at the ear ground. The
dotted region represents the concentrations (mean
values ± SD) in control calves.

52

Due to the latter fact it remains a problem to set up an
appropriate monitoring program to control treatments of
properly administered endogenous sex steroids. Hence also the
question to appoint a withholding period may be challenged,
provided that the implantation site (ear) will be removed at
slaughter. In a study that simulated an often used illegal
treatment we injected 50 mg oestradiol-benzoate intramuscu-
larly. After slaughtering the calf three weeks later the
tissue, apart from the injection site, contained oestrogens
within the level of the control calves (C in Table 7). Although
there was a higher content of total oestrogens in the
injection site (up to 57 ng/g compared to 7 pg/g in peripheral
muscle), the total amount of the still esterified steroid
in a piece of 1 kilogram contaminated meat adds up to
approximately 15 µg only (Table 8).

By monitoring this experiment (Table 9) we see a quick
clearance of the steroids within the first week. For example
from a female control animal values up to 39 ng total
oestrogens/g fresh faeces (≙ 158 ng/g dry faeces) were
obtained. Consequently, we can determine such an illegal
treatment only during the first days. On the other hand,
determinations of total oestrogens in faeces (Figure 10)
of properly treated animals also overlap with controls. In
consequence there are only rare hints for a treatment even if
a few higher values are obtained from treated animals (148 ng/g
fresh faeces ≙ 844 ng/g in dry faeces). The mean values are
generally higher in treated _female_ animals. In excreta of the
male veal calves (Figure 11) we found rather high oestrogen
concentrations even in control samples (urine, 27 ng/ml; 80 ng/g
fresh faeces ≙ 279 ng/g dry faeces). Elevations due to the
treatment are indicated only during the first four weeks. Due
to overlappings with control values during the course of the
investigation it seems doubtful to distinguish between treated
and untreated animals by a single hormone determination.

TABLE 8: Oestradiol-benzoate in pieces of muscle (24 to 74 g) next to the injection site three weeks after i.m. application of 50 mg oestradiol-benzoate in oily solution.

Distance to injection site/cm	concentration pg/g	total ng
− 30	34	3
− 25	64	5
− 20	50	1
− 15	6	0,1
− 14	106	3
− 13	18	1
− 12	24	1
− 11	5384	203
− 10	847	28
− 9	224	8
− 8	2855	111
− 7	30	1
− 6	11925	340
− 5	26545	851
− 4	24085	1020
− 3	30615	1155
− 2	21682	922
− 1	20498	894
injection site	56670	2170
+ 1	53065	2200
+ 2	31461	1380
+ 3	19491	748
+ 4	26238	1066
+ 5	16307	604
+ 6	8425	461
+ 7	7229	270
+ 8	4175	165
+ 9	577	14
	summarized	14.6 µg E_2

TABLE 9: Measurement of total oestrogens (free and conjugated)
after i.m. injection of 50 mg oestradiol-benzoate.

Day	faeces ng/g	urine ng/ml	plasma pg/ml
0	6	6	160
+ 1	56	10600	not measured
+ 2	20300	500	1980[x)]
+ 7	59	16	not measured
+14	211	10	not measured
+21	33	15	260

[x)] further 18 ng esterified oestradiol per ml
plasma were found.

In Figure 12 the results for testosterone in plasma are presented. In the female calf the small elevation up to 21 days
after treatment is below the ranges of the normal testosterone
values in the male one. In the untreated male calf the rise of
testosterone during the course of approaching puberty and its
suppression in the treated one was obvious. The experiment
corresponds with the data presented in Figure 11. By shifting
the implantation site contrary to the prescription to the ear-
ground (investigation with female calves corresponding to
treatments A, Table 7) testosterone values were found to be
elevated during the first three weeks after administration.

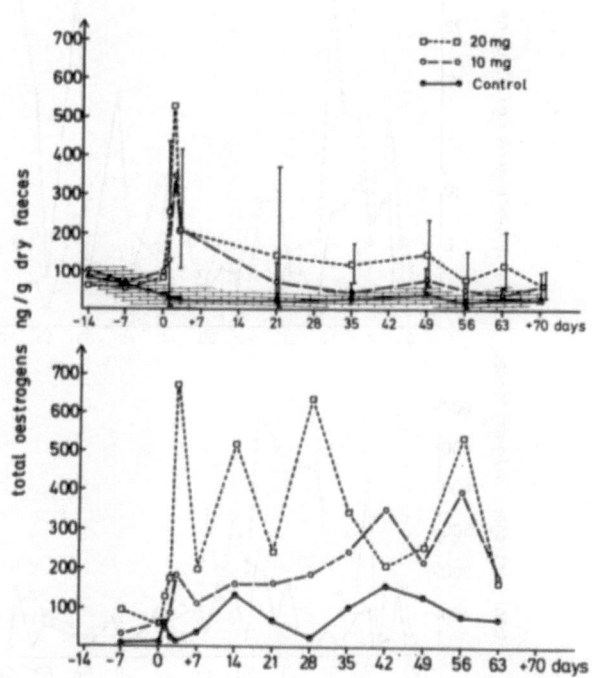

FIGURE 10: Total oestrogens in faeces after treatment of
female veal calves with oestradiol-benzoate in
combination with testosterone propionate.

Upper panel: each treatment n=6;
             implantation s.c. at the ear ground.
             One standard deviation in the
             control group is shown by the dotted
             region and in the treatment groups
             by solid bars.
Lower panel: each treatment n=1;
             implantation s.c. on the ear
             cartilage.

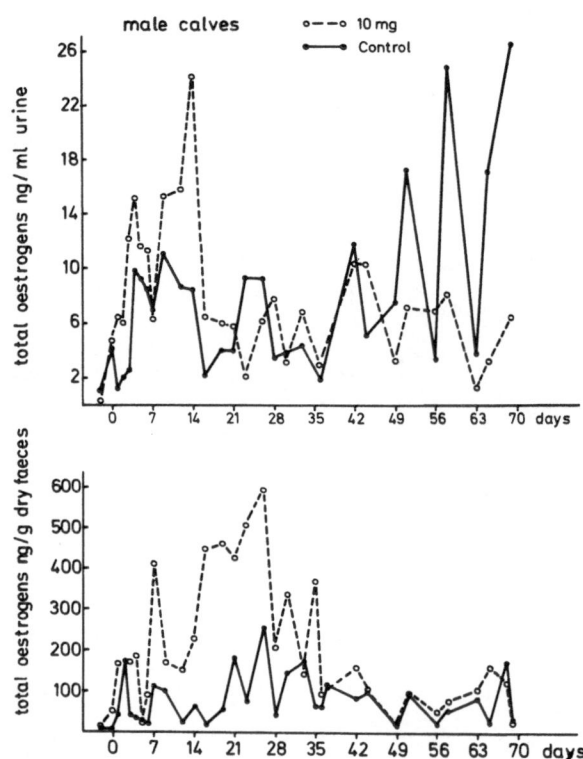

FIGURE 11: Excreted total oestrogens from treated (10 mg
            oestradiol-benzoate in combination with 100 mg
            testosterone-propionate s.c. on the ear cartilage)
            and untreated male calves (each treatment n=1).

            Upper panel: urine;
            lower panel: faeces.

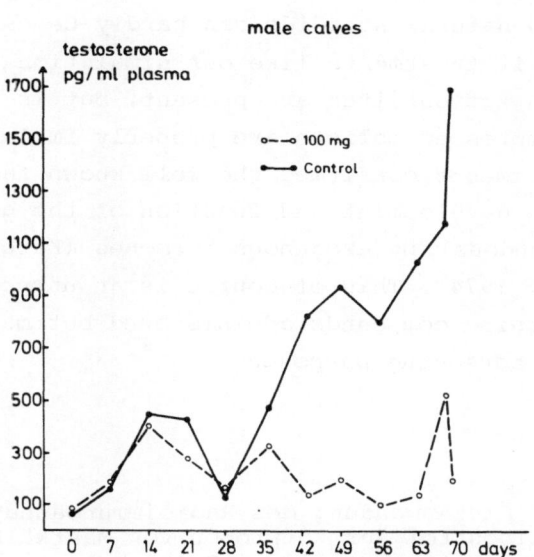

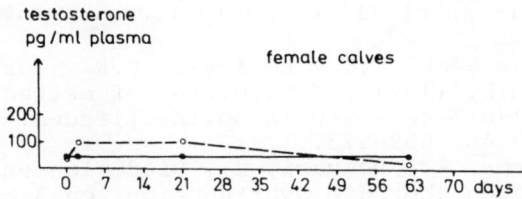

FIGURE 12: Total testosterone in plasma of veal calves after
implantation of 10 mg oestradiol-benzoate in
combination with 100 mg testosterone-propionate
s.c. on the ear cartilage.

Upper panel: each treatment n=1; male;
lower panel: each treatment n=1; female.

Finally, it must be concluded that a monitoring program for treatments with natural steroids can hardly be established. It is possible if treatments like our simulation study occur and critical injection sites are present, but is not possible if suitable compressed pellets are properly implanted. The described experiments confirmed the well known fact that in veal calves the development and function of the gonads are depressed tremendously by exogenous hormones (Karg et al.,1972; Grunert et al., 1974). This of course is an unspecific criterion concerning compounds administered but may be considered for screening purposes.

REFERENCES

Agthe, O. 1980. Die Anwendung des Radioimmunoassays für Diäthylstilböstrol auf Kotproben von Mastkälbern. Arch.f.Lebensmittelhyg. 31, 102-105.
Boursier, B., Richou-Bac, L. and Dezothez, C. 1982. Contrôle de l'administration de diethylstilbestrol chez le veau à l'abattoir possibilités analytiques. Rec.Méd.vét. 158, 315-320.
Dunn, T.G., Kaltenbach, C.C., Koritnik, D.R., Turner, D.L. and Nisender, G.D. (1977). Metabolites of estradiol-17ß and estradiol-17ß-3-benzoate in bovine tissues. J.Anim.Sci. 46, 659-673.
Grunert, E., Harms, F., Pozvari, M., Diederichsen, J. 1974. Untersuchungen über die Ovarfunktion von 3 bis 5 Monate alten Mastkälbern nach Östrogenzufuhr und ihre Bedeutung als Screening-Test. Arch.f.Lebensmittelhyg. 25, 188-190.
Henricks, D.M. and Torrence, A.K. 1977. Endogenous estrogens in bovine tissues. J.Anim.Sci. 46, 652-658.
Henricks, D.M. 1981. Assay of naturally occurring oestrogens in bovine tissues. In: Proc.Intern.Symp."Steroids in animal production", April 1980 in Warsaw. Ed. Henryk Jasiorowski, Warsaw Agricultural University, 161-170.
Huis in't Veld, L.G., Jonkman van den Broek, E.B., de Groot, W.C. and Cornel-Krüsel, H.C. 1969. The concentration of substances with oestrogenic activity in the musculature of the injection site and in urine of calves given diethylstilboestrol and hexoestrol. Neth. J. Vet. Sci. 2, 171-178.
Hoffmann, B. and Oettel, G. 1976. Radioimmunoassays for free and conjugated trienbolone and for trienbolone acetate in bovine tissue and plasma samples. Steroids 27, 509-523.

59

Hoffmann, B., Heinritzi, K.H., Kyrein, H.J., Oehrle, K.L.,
    Oettel, G., Rattenberger, E., Vogt, K. and Karg, H. 1976.
    Untersuchungen über Hormonkonzentrationen in Geweben,
    Plasma und Urin von Mastkälbern nach Behandlung mit
    hormonwirksamen Anabolika. In: Fortschritte i.d. Tier-
    physiologie und Tierernährung (Beiheft z.Ztschr. Tier-
    physiol. Tierernährg. u. Futtermittelkunde). Ed. J.
    Brüggemann and O. Richter (P.Parey, Hamburg-Berlin)
    6, 80-90.
Hoffmann, B. and Rattenberger, E. 1977. Testosterone
    concentrations in tissue from veal calves, bulls and
    heifers and in milk-samples. J.Anim.Sci. 46, 635-641.
Hoffmann, B. 1978. Use of Radioimmunoassay for monitoring
    hormonal residues in edible animal products.
    J. Assoc. Off. Anal. Chem. 61, 1263-1273.
Hoffmann, B. and Laschütza, W. 1980. Entwicklung eines
    Radioimmuntests zur Bestimmung von Diäthylstilböstrol in
    Blutplasma und eßbaren Geweben vom Rind. Arch.f.Lebens-
    mittelhyg. 31, 105-111.
Hoffmann, B. 1981. Aspects on metabolism, residue formation
    and toxicology of growth promoters. Arch.f.Lebens-
    mittelhyg. 32, 65-69.
Hoffmann, B. and Blietz, C. Application of radioimmunoassay
    (RIA) for the determination of residues of anabolic sex
    hormones. J.Anim.Sci. in press.
Karg, H., Hoffmann, B., Vogt, K. and Behr, H. 1972. Radio-
    immunologische versus fluorimetrische Bestimmung
    exogener und endogener Steroidöstrogene bei Mastkälbern.
    Tierärztl. Umschau 27, pp.385.
Meyer, H.D.   Neue Methode zur Bestimmung von Steroid-
    östrogenen im Gewebe. Wiener Tierärztl. Mschr. in press.
Oehrle, K.L., Vogt, K. and Hoffmann, B. 1975. Determination
    of trenbolone and trenbolone acetate by thin-layer-
    chromatography in combination with a fluorescence colour
    reaction. J. Chromatography 114, 244-246.
Ryan, J.J. and Hoffmann, B. 1978. Trenbolone acetate:
    Experiences with bound residues in cattle tissues.
    J.Assoc. Off. Anal. Chem. 61, 1274-1279.
Schopper, D. 1981. Bestimmung von Trenbolon (TBOH) im Urin und
    Kot von Mastkälbern mittels eines radioimmunologischen
    Verfahrens. Arch.f.Lebensmittelhyg. 32, 203-208.
Schopper, D. and Hoffmann, B. 1981. Identifizierung von 17a-
    Trenbolon als Hauptausscheidungsprodukt des Trenbolon-
    acetat-Stoffwechsels beim Kalb und sich daraus ergebende
    Konsequenzen für die Rückstandsanalytik. Arch.f.Lebens-
    mittelhyg. 32, 141-144.
Schopper, D., Hoffmann, B., Karg, H., Berende, P.L.M.,
    v. Weerden, E.J. and v. der Wal, P.
    Untersuchungen zur analytischen Erfaßbarkeit der Zufuhr
    von 17ß-Östradiol und Trenbolonacetat in Blutplasma und
    Urin beim Kalb. Manuscript in preparation (Zbl.Vet·. Med.)

60

Vogt, K., Waldschmidt, M. and Karg, H. 1970. Bestimmung von
Ausscheidungsverlauf und Rückständen von Diäthylstilb-
östrol nach intramuskulärer Verabreichung beim Kalb mit
Hilfe biologischer und chemischer Nachweisverfahren.
Berl.Münchn.Tierärztl.Wschr. $\underline{83}$, 457-462.

Vogt, K. 1977. Dünnschichtchromatographisch-fluorimetrische
Bestimmung von Trenbolon und 17a-Östradiol im Urin
männlicher Mastkälber nach subkutaner Verabreichung von
Revalor [R], Trenbolonacetat oder 17ß-Östradiol.
Arch.f.Lebensmittelhyg. $\underline{28}$, 177-180.

Vogt, K. 1978. Dünnschichtchromatographisch-fluorimetrische
Bestimmung von Diäthylstilböstrol in Kot und Urin von
Mastkälbern. Arch.f.Lebensmittelhyg. $\underline{29}$, 178-181.

Vogt, K. 1979. Weitere Verbesserung des Nachweises von Stil-
benderivaten mit Hilfe der dünnschichtchromatographisch-
fluorimetrischen "Dansylierungs"-Methode. Arch.f.Lebens-
mittelhyg. $\underline{30}$, 168-171.

Vogt, K. 1980. Vereinfachtes Extraktions- und Reinigungsverfah-
ren für die radioimmunologische Bestimmung von Diäthylstilböstrol in Fleisch, Leber und Niere. Arch.f.Lebens-
mittelhyg. $\underline{31}$, 138-141.

Vogt, K. and Karg, H. 1982. Modellversuch zur Beurteilung der
Rückstandssituation nach intramuskulärer Verabreichung
einer kristallinen Suspension von Diäthylstilböstrol-
dipropionat an Mastkälber. Zbl. Vet. Med. A $\underline{29}$, 279-288.

WHO Working Group 1982. Health aspects of residues of anabolics
in meat. Euro reports and studies $\underline{59}$, 1-38.

DISCUSSION ON PROFESSOR KARG'S PAPER

Dr. Verbeke: We did some experiments in steers where we injected DES in an oily solution. After 30 days we did not find any residues at the site of injection. We measured residues from different tissues as a function of time. In some tissues there were no residues up to 90 days but between 90 and 150 days we found some residues. Have you any explanation for this peculiar finding?

Prof. Karg: Were the animals under your own control?

Dr. Verbeke: Yes.

Prof. Karg: Was there a second injection given?

Dr. Verbeke: No. We followed levels in urine and these were low for 60 - 90 days post injection and then levels increased.

Prof. Karg: That may be a rare incidence of formation of a capsule at the site of injection.

Dr. Verbeke: No. The injection site of every animal was checked.

Prof. Karg: We have no such experience. There is a variation in output of hormone within a certain range but in general the former level is always higher than the latter.

Dr. Heitzman: Is it possible that like progesterone if you use rapidly absorbed injections there could be a 2nd depot site namely fat?

Prof. Karg: DES has no lipophilic properties like progesterone. Progesterone is cleared after 2 - 3 days.

GENERAL   DISCUSSION

Dr. Davis: Monoclonal antibody procedures lend themselves, theoretically at least, to generation of antibodies that are specific. Your report however, suggests that there was some non-specificity. Please comment. Secondly, comparitively what is the sensitivity of assays using monoclonal antibodies or in vivo produced antibodies?

Dr. Heitzman: Clones are selected on the basis of specificity not only to related metabolites but also to avoid high blank values. It depends on the number of clones examined. The antibody produced against zeranol was the best clone found following examination of about 100 clones. It has very good specificity. It is possible by examining more clones to get an antibody of even higher specificity depending on the time available. Specificity depends on how many clones you are prepared to examine. You could chose an aspecific antibody for certain use if this was desired. As far as sensitivity goes there is some suspicion that monoclonal antibodies are less sensitive and have a lower affinity. So far we have not found that to be true. The calibration curves with monoclonal antibodies are in the same region and give the same sensitivity as polyclonal antibodies.

Prof. Hoffmann: Monoclonal antibodies may not necessarily be better than polyclonal antibodies.

Dr. Heitzman: The only advantage of monoclonal antibodies is their permanancy i.e. once you have it you always have it. A rabbit has a finite lifetime.

Prof. Hoffmann: Specificity of antibodies is important but for the RIA, extraction procedures must preceed the assay.

Dr. Verbeke: It is not the absolute sensitivity which determines the result but interfering substances can give

rise to false positives. Have you any experience with
zeranol or with veal calves treated with DES?

Dr. Heitzman: Interference from non identifiable substances
is the biggest problem of all not only in RIA but in any
end point analytical system measuring in the picogram range.
The problems with zeranol initially were not due to the
antibody but to the high background interference which
is common to many assay systems. A minimum of extraction
is necessary to reduce background noise and minimise
labour costs.

Prof. Hoffmann: The sensitivity of the assay must be high
in order to allow a high level of accuracy of detection
and differentiation between a few picograms of compound.

Dr. Heitzman: The situation with DES is particularly
interesting and the main problem has been its use in the
veal calf. The limits of detection have been adequate
to allow detection of DES treated veal calves. But many more
animals in Europe have been implanted with DES, particularly
adult beef cattle. A high percentage of pigs in U.K. have
also been treated with DES. With the older assay methods used,
the limits of detection set were so high that residues
in DES or hexoestrol treated animals would never be found.

Dr. Verbeke: RIA is used for screening, mostly on urine
samples. If you check meat you do not find any residues.
Positive detection by RIA has to be confirmed by a
second independent method with the same sensitivity. With
the throughput of 300 - 400 samples per week, if there is
a high positive rate then you run into a problem with
throughput from the chemical method which is only 30 - 40
samples per week.

Dr. Heitzman: I don't think that is a major problem
because any country with one third of its samples positive

is in real trouble. From discussion with representatives
from member countries none or very few positive samples
are found with RIA for DES. The chemical method to be used
depends on the concentration and tissue used e.g. the
concentration is generally higher in excreta than in
edible tissue.

Prof. Karg: The required sensitivity of the assay methods
is very different for urine, kidney or muscle. Detection
in muscle outside the injection site requires a very
sensitive RIA. Use of muscle alone makes it very
difficult to detect illegal injections. However detection
of such illegal injections is aided by monitoring liver,
bile, urine and faeces.

Prof. Hoffmann: In relation to detection of illegal
use of DES, it is much easier to detect this by using urine
or faeces compared to tissue samples. This has proved
effective and allows faeces from the live animals to be checked. In 1980, 40% of the veal calves were found positive
using the faeces method. Within six months the
incidence in Germany dropped to less than 1% and two weeks
ago it was reported that no DES positive calves
were found in Germany in spite of the fact that hundreds
of samples are checked using highly sensitive methods. But
we still require a method for tissue to monitor carcases
imported into Germany.

Dr. Verbeke: Monitoring of the DES-residues in the urine
of cattle allow a screening of the treated animals "in vivo".
Amongst the methods reported for the detection of DES,
RIA -analyses are fast and adapted for screening a large
number of samples. However, while implementing statutory
control measures in the Netherlands, RIA-analyses were observed to yield a large number of false DES-positives, stressing
the need for a fast, sensitive and reliable chemical method
to confirm results obtained by RIA-analyses. We report

here a new fast routine analytical procedure allowing for a highly specific quantitative determination of DES below the ppb-level (Verbeke and Vanhee 1983, submitted to J. Chromatogr.)

The method is based on a simple, specific absorption of oestrogens from a urine hydrolysate on a pyrrolidone-resin and on an in-line specific conversion of DES by successive irradiation and reduction to highly fluorescent products (Banes, D., J. Ass. Off. Agr. Chem. 1961, 44, 323 - 328) after HPTLC-chromatography. Fluorescence spectroscopy at optimised excitation and emission wave-lengths allows quantification of a few picogram DES in the column-effluent.

Analyses of more than 60 samples containing O.5 - 5 ppb DES yielded results which correlated closely with those obtained by an independant quantitative TLC-fluorescence procedure (Verbeke, R., J. Chromatogr. 1979, 177, 69 - 84). The procedure is rapid: 15 samples can be extracted and analysed in one working day. This work was supported by grants from the Belgian I.W.O.N.L.-Foundation.

Prof. Hoffmann: I would like to stress that EEC veterinary guidelines require drug monitoring for all veterinary drugs, so we are faced not only with the problem of anabolics but also antibiotics, pesticides, anthelmintics etc. It would be nice to have a multi-residues analysis but this may not be feasible for some countries who may just specifically determine the presence or absence of a compound. It would be good for those countries not having sophisticated equipment, if there were more simplified techniques available. Can Dr. Meyer comment on the technique he has been using.

Dr. Meyer: For the radioimmunological determination of hormonal steroids extracted from tissues and excreta, prior purification is necessary. Besides the classical

66

liquid-liquid partition methods, nowadays HPLC on RP-18
Silicagel most of the UV-absorbing material does not
migrate with the steroids, its separation seemed to be
easy. A rapid purification procedure has been developed
using Sep-Pak C$_{18}$ cartridges. The cartridges are plugged
to a syringe and the aqueous sample (containing less than
30% methanol) is pumped through the cartridge. In case
of oestradiol the cartridges are washed with 5ml. of 40%
methanol/60% water·and oestradiol is eluted with 2ml.
100% methanol. For other steroids the conditions are somewhat different. 90% of the UV absorbing material is
separated from the steroids by this rapid and very
economical method. With pure solvents procedural blanks
can totally be avoided and steroid determinations in the
lowest picogram range are possible. This sample clean up
in Sep-Pak C$_{18}$ cartridges prior to radioimmunoassay will
be published in Acta endocrinologica (Kbh.) Suppl. March
1983.

Prof. Hoffmann: A whole array of methods is available.
It is most important that the end point determination is
done properly and the correct answer is obtained. There
are two methods which so far have been used most
successfully for monitoring anabolic residues. The
historical approach was thin layer chromatography which was
quite successful for urine and faeces and more recently,
in the last 3 - 4 years, RIA. The important question is
do we measure the right compound in these assays in
respect to consumer toxicology. This brings us to
metabolism. Are there any questions with respect to the
metabolism of these compounds from the papers presented
this morning.

Dr. Meyer: If trenbolone 17α is epimerised in the liver
and blood are you really sure that trendione is the only
substrate for this epimerisation because generally for
other epimerises both the reduced and the oxidised
substrate can be used.

Dr. Heitzman: No, I am not sure. So far we have been
unable to convert the trenbolone β form to the α form in blood.
This was in vitro. We have also been unable to isolate
from the liver any enzyme which will make 17α either from β
or starting with the ketone and yet we have no problem
in isolating the enzyme in the liver that converts the β to
the ketone. I don't think we have the final answer.
It is likely that there is some minor enzymatic activity and
also most of our work has been done in vitro with liver
homogenates or enzyme preparations or taking whole red
blood cells from the animal. We have done quite a lot
of new work where we have treated animals in vivo and that
is not complete yet. But we find almost 100% of 17α metabolites
in the bile and we find 17β and 17α circulating in the blood
both free and conjugated. We have animals cannulated
across the liver so that we can measure the conversion
between the blood that enters the liver and the blood-bile
which leaves the liver. The metabolic transformations that
take place within the liver are difficult to determine
because, unfortunately, of the rather large blood volume
within the liver. So it is a very difficult problem and I
would like to hear what Professor Rico has to say about it
this afternoon.

Dr. Verbeke: Have you ever tried to give to cattle
excessive doses of natural hormone, and if so after 3 weeks
have you seen higher or normal levels?

Prof. Karg: We have not done it in adult cattle. In a calf
treated with 50mg. oestradiol benzoate in oily solution into
the muscle, three weeks later we found normal levels in all
edible parts of the body except at the injection site. There
were levels in the nanogram concentrations there. It was
calculated that the total concentration at the whole injection
site in muscle was about 15 micrograms.

Prof. Hoffmann: Dr. Karg, you showed that testosterone in
vivo was depressed in animals treated with testosterone

68

propionate implants compared with control animals. Would
you not expect lower tissue levels of testosterone in these
animals. Do you have any comparison?

Prof. Karg: We have only measured plasma testosterone
but not exact tissue testosterone, but Dr. Meyer
may wish to comment.

Dr. Meyer: Unfortunately we have not measured testosterone,
but in the case of oestradiol,it was higher in faeces,
urine, plasma and other tissues examined,of the control
animal.

Prof. Hoffmann: So you could conclude that treatment of
intact animals with anabolic agents reduces endogenous
steroids.

Prof. Karg: It also delays puberty.

Dr. Schanbacker: The problems as I see it are ones of
sensitivity and specificity of assays. I would like to
express concern about in vitro and in vivo assays i.e.
immuno and bio-assays. I would like to suggest that
if we define the metabolites of most concern we can produce
antibodies against them e.g. against the α and β metabolites
with high specificities as Dr. Heitzman suggests. With
blood samples and new techniques we are running 5000 assays per
week for testosterone in our laboratory. It is possible to
push through a lot of samples and do the screening and
monitoring studies, and perhaps that will answer some of
the questions in regard to monitoring.

Dr. Heitzman: If tolerances are set for trenbolone and
zeranol, I think they are likely to be fairly high and easy
to measure by almost any method.This will allow us to get
a very sensitive method and use it on a small number of
samples. In our use for monitoring, it will probably have
to be urine or bile. We can carry out extremely fast

assays using very specific antibodies with small amounts of
material, because the background is now very low. This
will allow large scale throughput and screening of
tolerance levels. That is fine until we come to edible
tissue. Unfortunately what we eat is muscle, liver and kidney,
and there will still have to be some kind of extraction
procedure for these. It may be possible for some tissues
to squeeze out the exudate for assay e.g. the drip from
liver. This should reduce the cost of the assay considerably.

Prof. Hoffmann: It would be inappropriate to speculate
how high the tolerance levels should be set. This is still
an open question. It may be excessively high or it may
be very low. I simply don't know.

Dr. Heitzman: If they are high it will make the methodology
much easier.

Prof. Hoffmann: There is no doubt about that. Any comments
on tolerance levels?

Prof. Van der Waal: I think the report of the Scientific
Expert Group is highly stimulating. A concensus among
a group of four different committees about what has been
achieved and what is missing is important. What is missing
requires an extensive research programme. It is important
to concentrate on the anabolic steroids as regards the proper
application, withdrawal time, dose etc. To proceed with
that there are two options -
(i) try and agree on a common programme among scientists
to reach a satisfactory outcome and later on to check whether
such a programme is sufficient, (ii) create a common
platform where we can try and formulate a tentative concensus
on what should be done. It seems a very urgent matter
because any programme will be time consuming and costly.
I would plead for a concensus within this group and within
the EEC Scientific Committee in order to be in harmony

within the Community. We should also try and co-ordinate efforts with other groups such as the FDA in the US.

Prof. Hoffmann: There will always be national problems. I don't think national governments can be overruled so easily at present. If the countries within the EEC can get along it is a big achievement. The problems mainly concerning toxicology and safety must be answered first. If we can come to conclusions there, then other questions can come up - how could a compound be safely used, what is the best monitoring programme to use etc. There is discussion between FDA and the EEC to try and find a common agreement to these problems. But a complete solution cannot be given at present. It is not a national problem. It is up to those commercial companies that want to market these compounds to supply adequate information and to update their information to comply with the latest EEC regulations that have been published. This information regarding the kind of toxicological information that should be presented by a company in order to get a compound accepted, is available to everybody.

Prof. Van der Waal: To design programmes to answer these questions, it is of use to have a co-ordinated approach. How is this to be achieved?

Prof. Hoffmann: From what the EEC has said the situation will be kept under review until 1984.

Dr. Raymond: (i) How will the Commission advise the Council and therefore individual countries as to the use of exogenous agents in the period between now and when this new information is available. Are they going to make a temporary recommendation on this? (ii) How will the Commission convince the public that highly valid scientific evidence on hormones is safe?. Very soon we will have to start a public education campaign to accept scientific conclusions.

Prof. Hoffmann: EEC has made a decision that in those countries where the xenobiotics are used, this situation will
not change until 1984. But then if enough information is
available on which to make a common decision, all EEC
member states will have to follow this recommendation.
There are rigorous safety criteria which all veterinary
products must go through before being let onto the market.

NEW DATA ON METABOLISM OF ANABOLIC AGENTS

A. G. Rico[+] and V. Burgat Sacaze[++]

[+]Laboratoire de Radioéléments et d'Etudes Métaboliques
                                            (I.N.R.A.)
[++]Laboratoire de Pharmacie Toxicologie - Ecole Nationale
                                    Vétérinaire -
31076 Toulouse Cedex, France.

ABSTRACT
                            The exogenous anabolic
agents, such as trenbolone acetate and zeranol have
variable   activity   by the oral route and are less readily
metabolized in the liver than the endogenous anabolics.
Their metabolic pathways can be complex and lead to excreted
forms after glucurono conjugation. With respect to biochemical mechanism of action, it can be assumed that the
anabolics act like all steroids by way of intracellular
receptors. Biotransformations could lead to molecules
with higher activity that may bind to normal constituents
of the organism. Bound metabolites are in general formed
later than free metabolites, and are considered less
harmful with a lower level of bioavailability. Recent
results concerning the affinity of anabolics for D.N.A.
seem to exclude a genotoxic action and confirm their
epigenic activity as potential carcinogens.

INTRODUCTION

Over the past decade, there has been much refinement in
the metabolic study of drugs. This trend is due to progress
in analytical methods that have become both more effective
and more reliable. The use of radio-labelled molecules
(Rico, 1976 ; Rico et al., 1980) in conjunction with high
performance liquid chromatography, gas chromatography,
nuclear magnetic resonance and mass spectrometry has led to
remarkable results.

The purpose of these metabolic studies is to obtain
information on the distribution, biotransformation, storage
organs, pathways and rates of excretion of drugs being
tested, and when used in food animals to determine accurate
withdrawal periods. Additionally, these metabolic studies
also provide more extensive data on the biochemical mechanism
and toxicity of these drugs, for both the target species and
the consumer of foodstuffs from treated food animals. The
purpose of this article is to review briefly these aspects.

In the first part, the authors shall discuss the metabolic aspects relative to the exogenous anabolics. In the second part, the problem of bound residues and in the last part some recent approaches concerning the possible carcinogenic risks of these drugs will be discussed. The metabolic pathways of endogenous anabolic agents will not be evoked because scientists now think that no question of safety arise in relation to the proper use of 17β-oestradiol, progesterone and testosterone in an appropriate form of preparation.

METABOLIC ASPECTS RELATIVE TO THE EXOGENOUS ANABOLICS

The exogenous anabolics in question are trenbolone acetate and zeranol which belong to the androgen and oestrogen groups, respectively.

Trenbolone (17 β-hydroxy trenbolone)

Trenbolone is a synthetic steroid with similar androgenic activity to testosterone but with greater anabolic activity. It has several structural features similar to testosterone but unlike testosterone, it has three conjugated double bonds. The material normally used in commerce is trenbolone acetate (T.B.A.).

The metabolism of trenbolone acetate has been studied in the rat, calf, heifer, cow and steer by Pottier (1978). Trenbolone acetate is rapidly hydrolysed to trenbolone, and both trenbolone and its metabolites are rapidly excreted as the glucuronides and sulphates, mostly in the bile. The predominant metabolites identified in the extractable fractions of bile in the rat are trenbolone and a 16-OH and a 17-keto metabolite ; while in cattle it is mainly 17α-hydroxy trenbolone with small amounts of trenbolone and other metabolites. Many of the minor metabolites have been identified in both species. These biotransformations explain the relatively weak activity of T.B.A. by the oral route.

Following the recommended use of trenbolone acetate in cattle and the application of various methods of assay (radioimmunoassay, thin-layer and high performance liquid chromatography) the main metabolite found in muscle has been shown to be trenbolone ($<$ 0.5 ppb muscle) and in liver and kidney a conjugated form of 17α -hydroxy trenbolone ($<$3 ppb). 17α -hydroxy trenbolone has a biological activity at least 10 times less than hydroxy trenbolone (Raynaud 1970).

Similarly, following subcutaneous implantation of cattle with trenbolone, radio-labelled with tritium, total residues, expressed as trenbolone equivalents, were a few parts per billion (10^{-9}) in muscle and approximately 10 times higher in liver and kidney (5-35 ppb) (Ross, 1980). 5-10% of the total residues labelled with tritium are extractable from meat, liver or kidney, and have been identified as similar to those found in bile. The identity of the remaining unextractable residues, labelled with tritium, which are most likely irreversibly bound, is not known but some tritium is found in the body water (Ryan and Hoffmann, 1978). The aspect of bound metabolites will be discussed in the second part.

Zeranol

This substance is structurally not a steroid and does not itself occur in nature although the parent substance zearalenone is a natural substance elaborated by Gibberella zea. It has oestrogenic and anabolic activity. The metabolism of this substance is less well known than that of trenbolone. It has been studied in a number of species including cattle and sheep. Most of the orally administered substance is absorbed and elimination is species dependent. Zeranol is excreted both as free and conjugated substance and as free and conjugated zeralanone (Migdalof, 1982). The same metabolites appear in the bile. Most of orally adminis- tered zeranol has practically disappeared from the tissues within 24 hours except in the case of liver and kidney. A recent, still unpublished work, reports that values found in the tissues are less than 2 ppb, 45 days after

implantation of 36 mg of zeranol in calves. There is no
accumulation in the organs. A great deal is still unknown
about the proportions of free and bound metabolites and
their exact nature.

The mechanism of action of anabolic agents.

Since the works of O'Malley and E.E. Baulieu (1978), it
has been clearly demonstrated that steroid hormones bind
to intra-cellular receptors within the cell. The crossing
of the cell membrane is due to facilitated diffusion. When
there are no hormones in the cell, the receptor is located
in the cytoplasm. Following binding of steroid hormone
an "active form" of the receptor seems to be stabilized
and it is transferred into the nucleus, where it attaches
itself to the chromatin and modifies the genetic expression
of the cell. A rather early increase in the synthesis of
proteins and the concentration of the mRNA (s) is observed.
The problem is to know if such a mechanism can be elicited
for the action of anabolic drugs. Even though there are
few receptors in muscle tissue, it cannot be excluded
that muscular protein biosynthesis could be affected in this
manner, especially with androgenic anabolics.

In any case, it has been established that anabolics
have affinities for hormone receptors; e.g., DES and zeranol
bind to the oestrogenic receptors of the uterus of the
bovine or human, and trenbolone binds to the androgenic
receptors. This binding may be the cause of the expression
of hormonal side effects that can occur after ingestion by
humans. It is important to keep in mind that receptors are
cytosolic proteins of high affinity, low capacity and
narrow biostereospecificity so that the hormone-receptor
binding is perfectly reversible. This may not be the case
for covalent associations that certain metabolites can have
with cellular constituents.

It appears from these characteristics that metabolism,
which changes the structure of the drug, decreases the
hormonal activity. Good examples are the difference in
biological activity which exists between trenbolone and

$\measuredangle$ -trenbolone and also between zeranol and zeralanone.

BOUND RESIDUES AND COVALENT BINDING

Bound residues can be defined as the non extractable radioactivity which persists in tissues after metabolic studies conduced with radiometric techniques.

In fact this fraction is found to contain:

(i) -"endogenous compounds", natural, resulting from incorporation of the degradation products of the parent compound at the intermediate stage of metabolism (amino acids, proteins, lipids etc.).

(ii) -"new compounds" resulting from covalent binding of the parent drug or metabolites to endogenous macromolecules.

It is evident that the first fraction is not toxic. Concerning the second, it is important to discuss it. Covalent binding is the covalent association of a reactive metabolite and an endogenous molecule. Through oxidation, some drugs can lead to the formation of electrophilic metabolites that can react with the nucleophilic groups in some cellular constituents: the amine, hydroxyl or thiol groups of DNA, RNA and proteins. This leads to the formation of covalent complexes. These reactions are unspecific, irreversible and imply, except for direct carcinogens, a bioactivation giving a short-lived and highly reactive metabolite. These characteristics are very important, because they indicate that the process of covalent binding is thermodynamically irreversible and the direct consequence is that even if we have cleavage of bonds by biological action such as hydrolysis, the released compound will never be the active metabolite. A good example is given by bromobenzene which can bind with proteins by 3-4 epoxide, ultimate negrogenic agent. These covalent complex macromolecules can, of course, be broken down and no longer yield the toxic epoxide. This conclusion has been confirmed recently by a publication of J. Fox (1982) on the in vitro reaction between ellagic acid, a natural phenolic compound, and benzo(a)pyrene, a potent carcinogen. Ellagic acid gives adduct with benzo(a)pyrene. This adduct by hydrolysis gives free ellagic acid and tetraol derivatives of benzo(a)-pyrene which are unable to cause mutations or to act as carcinogens.

This concept of metabolites attached by covalent binding to normal cellular constituents is worthy of discussion. In fact, it leads to the differentiation of two types of compounds that may be found in foodstuffs from treated animals (V. Burgat-Sacaze, P. Delatour and A.G. Rico, 1981). The molecule itself and closely related metabolites of the slightly biotransformed initial molecule, in general occur early and do not accumulate because they are often quickly excreted. With regard to anabolics, mention is to be made of oestradiol-17β, oestrone, oestradiol-17α, DES and the conjugated substances of w-hydroxydienestrol, zeranol, TBOH, etc... These substances certainly still have some hormonal properties due to their potential binding with the specific receptors of the target organs. They could still in some cases be metabolized to highly reactive electro-philic intermediaries.

The bound metabolites can accumulate over time, in particular, in the case of repeated administrations or the implantation of a drug because excretion is slower. They represent a possible proof of aggression against the target species. Bound metabolites have lost their reactive potential and it can be assumed that they do not constitute a high level of toxicity for the consumer, particularly because their bioavailability is often low.

This theoretical concept has been confirmed by Jaggi et al. (1980). They demonstrated that the carcinogenic risk of humans who consumed liver or meat containing aflatoxin B_1-bound residues was negligible when compared with the risk from the intake of free substances.

The work of Ryan and Hoffmann (1978) on trenbolone discussed above also illustrates these important metabolic considerations. It was established by these authors that some of the non extractable residues by organic solvants

was water soluble and more was insoluble in water and
presumably bound to tissues. Concerning the insoluble
fraction, it was demonstrated that by the action of
different proteolytic enzymes such as pepsin and trypsin
the major part of these residues became water soluble.
These results indicate that, for the most part, bound
residues can change to highly polar compounds very different
to the parent drug. · This increase in polarity is a very
good indication of detoxification. These polar compounds
cannot have hormonal activity and have very low if any
reactive potentiality. This high polarity also certainly
explains their poor bioavailability. All this information
points to the idea that bound residues present very low
if any toxicity.

CARCINOGENIC RISKS

The existence of bound residues can be a proof of
possible genotoxicity of drugs. In these conditions, it
is necessary to study this kind of toxicity. The possible
genotoxic effects must be explored in different ways.

The mutagenicity of the parent drug and of the primary
metabolites provides useful information. Neither trenbolone
norα-trenbolone showed any mutagenic activity in prokaryotic
systems nor in any in vivo test for bone marrow and germ
cell cytogenetics. Clastogenic activity in vitro against
human lymphocytes in culture were negative but tests
against cells (L 5178 mouse lymphoma) were slightly positive,
but to the same degree shown by testosterone and oestradiol
(Ashby, 1981). Another possibility is the research of the
capacity for DNA binding. This capacity can be evaluated
by Covalent Binding Index (CBI) which is defined as μmoles
of bound product per mole of nucleotide divided by mmole
of administered product per kg of body weight per animal.
Lutz (1979) classified carcinogenic agents according to
their CBI values: strong carcinogens such as aflatoxins
present CBI in the order of thousands, average carcinogens
as A.A.F. present CBI in the order of hundreds; no or weak

carcinogens present CBI in the order of ten.

Preliminary results obtained, but not yet published by Dirheimer (1982) indicate that in vivo in rats the CBI of some anabolics is very low. These results are summarized in Table 1.

TABLE I. CBI OF ANABOLIC AGENTS.

	OESTRADIOL	TESTOSTERONE	T.B.A.	ZERANOL
Doses µg/kg	15	19.8	17.8	30
C.B.I.	11.4	4.8	5.6	8.5
Number of nucleotides / 1 molecule of hormone	1.66×10^9	3×10^9	3.2×10^9	1.25×10^9

It is important to stress, that the doses administered to rats, in the order of 20 µg/kg is in fact about four thousand times higher than the possible doses ingested by consumers of meat from treated animals. Since Lutz (1979) indicated that there is a linear dose binding relationship for CBI, it appears that in these conditions, the number of molecules of anabolics fixed on DNA would be four thousand times lower than the number found, which is in the order of one molecule for five to thirteen thousand trillion nucleotides. If we consider Watson (1973), that a gene constitutes in the order of three thousand nucleotides, the present results indicate that the action of anabolics could only change one gene per trillion. These appear as very negligible and exclude a genotoxic action of anabolic agents. Their eventual experimental carcinogenic effects can be explained by epigenic effects and by their hormonal side effects. Then it is possible to envisage for them

a hormonal treshold and that is why it is very important
to define their no hormonal effect levels.

CONCLUSIONS

In conclusion, I would like to stress some points.
The exogenous anabolics can have variable activity by
the oral route depending on the degree of metabolism.
When withdrawal periods are respected, the residual tissue
levels are always very low.

A more refined approach to metabolic studies, made
possible by developments in analytical methods, has brought
forth such new concepts as bioactivation and the covalent
binding of active metabolites to endogenous molecules. It
is thus essential to examine more closely the toxicological
impact on the consumer of various metabolites encountered
in the tissues of treated animals. The bound metabolites
are not likely to constitute a high toxicity risk for man.

From recent results concerning affinity for DNA,
trenbolone and zeranol appear as non genotoxic compounds.
Thus, their eventual carcinogenic effects can be explained
by their hormonal properties. Considering this second
remark, further metabolic studies will be necessary,
particularly with regard to exogenous anabolic agents. Such
advanced studies will lead to a better understanding of their
mechanism of action and any toxic effects.

It appears we are entering a new era of safety evaluation
and, like Golberg (1979), we would say that: "In the new
era of toxicology, it is essential to apply whatever tools
are appropriate to the study of a particular compound;
and if we do all this in an intelligent and well informed
manner, the new era will indeed have arrived. The public
and the environment will be the beneficiaries of our
objective and conscientious scientific efforts".

REFERENCES

Ashby. Genotoxicity and in vitro effects of hormones.
    Europ. Toxic. Forum, 1982, May 3-7, 339-346.

Baulieu E.E.. Les hormones. In : Hormones, aspects
    fondamentaux et physiopathologiques. Beaulieu, E.E.,
    Hermann éditeurs des sciences et des arts, 1978.

Burgat-Sacaze, V., Delatour, P. and Rico, A.G. Bound residues
    of veterinary drugs. Ann. Rech. vét., 1981, 13, (3),
    277-289.

Dirheimer, G. In vivo convalent binding to rat liver DNA
    of trenbolone as compared to 17β -oestradiol on
    trenbolone and zeranol. 1972, report unpublished.

Fox, J.. Compound neutralizes carcinogen in vitro.
    Chemical and engineering news, 1982, oct. 25, 26.

Golberg L. Toxicology : Has a new era dawned. Pharmacol.
    Rev., 1979, 30, 351.

Jaggi, W., Lutz, W.K., Luthy, J., Zweifel, U. and Schlatter,
    C.H. In vivo convalent binding of aflatoxin metabolites
    isolated from animal tissue to rat liver DNA. Food
    Cosmet. Toxicol., 1980, 18, 257.

Lutz, W.K.. In vivo convalent binding of organic chemicals
    to DNA as a quantitative indicator in the process of
    chemical carcinogenesis. Mutation Research, 1979,
    65, 289-356.

Migdalof, B.H., Dugger, H.A., Heider, J.G. and Coombs, R.A..
    Biotransformation of zeranol I. Disposition and metabo-
    lism in the female rat, rabbit, dog, monkey and human.
    Xenobiotica, 1982, accepted for publication.

Pottier, J. Report for registration unpublished. Centre
    Recherches Roussel UCLAF, Romainville, 1978.

Raynaud, J.P. Exc. Med. Int. Congr., 1970, serié n°219,
    915-922.

Rico, A.G. Contribution des techniques radioisotopiques
    à l'étude du métabolisme des médicaments et à la
    détermination des niveaux des résidus dana les tissus
    des animaux. Ann. Rech. Vet., 1976, 7, 161.

Rico, A.G., Bénard, P., Braun, J.P. and Burgat-Sacaze,
    V. Whole body autoradiography in metabolic studies
    of drugs and toxicants. Adv. Vet. Sci. Comp. Med.,
    1980, 24, 291.

Ross, D. B. Toxicology and residues of trenbolone acetate
    as a model. Communication : Steroids in animal
    production. Intern. Symp. Varsovie, 1980.

Ryan, J.J. and Hoffman, B. Trenbolone acetate : experiences
    with bound residues in cattle tissues. J. Assoc.
    Official Anal. Chem., 1978, 61, 1274.

Watson, J.D. Biologie moléculaire du gène. Seconde édition,
    Inter European Editions, 1973, 272-282.

DISCUSSION ON PROFESSOR RICO'S PAPER

Dr. Meyer: What is the nature of the protein that trenbolone is binding to?

Prof. Rico: We do not know the exact nature of the binding protein.

Dr. Meyer: We have carried out cellular fractionation and the radioactivity is located mainly on microsomes and on the cytosol protein.

Prof. Rico: It is not easy to characterize the exact binding proteins. Have you information on methodology to do this work?

Dr. Meyer: It should be easy as the proteins are already labelled.

Prof. Rico: Yes, but extraction is very difficult in order to study their exact nature.

Dr. Buttery: Where is the trenbolone molecule labelled?

Prof. Rico: In the steroid ring between C_5 and C_6.

Dr. Heitzman: Is the anabolic activity of αtrenbolone 5%?

Prof. Rico: Yes, from data in the rat.

Dr. Heitzman: Is there any evidence of higher anabolic activity of αtrenbolone in ruminants?

Prof. Rico: I can give you the reference where you can get this information.

Prof. Hoffmann: Can you directly relate the anabolic activity to the hormonal activity?

<u>Prof. Rico:</u> No. There are two types of potencies viz. androgenic and anabolic potency.

RECENT STUDIES WITH ANABOLIC AGENTS IN

STEERS AND BULLS

M. O'Lamhna and J. F. Roche,

University College Dublin,

Veterinary Field Station,

Ballycoolin Road, Finglas,

Dublin 11, Ireland.

ABSTRACT

In trial 1, 490 yearling steers were allocated at
random to the following treatments (a) control (b) given a
silastic rubber oestradiol implant (c) as in (b) but given
three implants of trenbolone acetate also (d) given three
implants of zeranol (e) as in (d) but given trenbolone
acetate at the same times. Oestradiol was given at the start
of the trial and zeranol and trenbolone acetate were given
at the start and on days 128 and 217 of the trial. The
daily liveweight gain for animals on each treatment were
0.69, 0.75, 0.78, 0.83 and 0.86 kg (SEM ± 0.02) respectively.
Carcass weights adjusted for days to slaughter for the five
treatments respectively were 300, 306, 311, 316 and 321 kg
(SEM ± 3.4). In trial 2, 133 Friesian bull calves, three
months of age, were allocated at random to the following
treatments (a) control (b) zeranol (c) 20 mg oestradiol and
140 mg trenbolone acetate (d) 20mg oestradiol and 200 mg
progesterone (e) silastic rubber implant containing 45 mg
oestradiol. The silastic rubber oestradiol implant was
given at the start of the trial, and the other pellet type
implants were given repeatedly every 103 - 115 days of the
trial. The daily liveweight gains for the 5 treatments were
0.71, 0.79, 0.81, 0.79 and 0.74 kg (SEM ± 0.02) and carcass
weights were 208, 215, 224, 225 and 209 kg (SEM ± 5.0)
respectively. Repeated use of pellet type implants increased
daily gain, and carcass weight and reduced testicular
weight and mounting behaviour.

INTRODUCTION

Anabolic agents, when administered as recommended
will increase growth rate, feed conversion efficiency and
lean meat in the carcass of beef cattle and veal calves
(Galbraith & Topps 1981). Studies from Ireland were
reviewed at the preceeding CEC meeting (Roche et al 1981)
and the aim of this paper is to update more recent information
relating to their use in bulls and steers.

One problem with the pellet type implants presently available is their short life span of 3-4 months (Roche & Davis 1983). Recently a silastic rubber implant containing oestradiol (Compudose, Elanco Ltd. UK) has become available with an expected life span of one year, which increases growth rate in steers (Mies et al, 1979). Trials were initiated to compare the effectiveness of this implant to repeated use of zeranol (Ralgro, IMC Ltd. US) with or without concommitant administration of the synthetic androgen, trenbolone acetate (Finaplix, Roussel Ltd France). The effectiveness of anabolic agents in bulls for beef production is not clear at present. It may be possible to reduce LH levels, delay puberty and reduce agressiveness and behavioural problems in bulls by utilising the negative feed back effects of oestradiol on LH secretion in cattle (Convey et al 1977). Trials were also carried out to determine the effects of repeated implantation of bulls with zeranol or oestradiol-progesterone combination on rate of gain, testicular weight and behaviour.

MATERIAL AND METHODS

Trial 1 : Friesian steers (n = 490) 12 - 15 months of age on 8 commercial farms were used. The animals on each farm were allocated at random to one of the following treatments:

(a) Control

(b) Implanted with 45 mg. oestradiol-17β (C-O) at start of experiment.

(c) Implanted with C-O and 300 mg. trenbolone acetate (TBA) at start and re-implanted with TBA on day 128 and day 217.

(d) Implanted with 36 mg. zeranol (Z) at start, on day 128 and day 217.

(e) Implanted with Z and TBA at start, on day 128 and day 217.

Implants were inserted into the ear using the appropriate recommended procedures. There were approximately equal numbers of animals per treatment on each farm. The animals were weighed at the start and at mean intervals of 55 days throughout the experiment. The first implant was given

in mid-April when the animals were at pasture, the second
in mid-August and the third at housing in mid-November.
Animals were housed in slatted floor buildings on six farms
and in open houses with straw bedding on the remaining two
farms. All animals were fed silage and concentrates while
housed. Carcass weights and conformation and fat scores
were obtained for 309 animals. Conformation and amount of
fat in the carcass were assessed by an official grader
using the Department of Agriculture's Beef Carcass Classification System. On 2 farms attempts were made to recover
all of the ears of the C-0 animals at slaughter in order
to determine the number of implants in site. The ear
inclusive of the base was removed and examined by blunt
dissection for presence of C-0 implants.

Trial 2 : Friesian bull calves (n = 133) approximately
three months of age were used. Animals were weighed at
the start of the experiment and allocated at random to
treatments as follows:

(a) Control
(b) Repeatedly implanted with 36 mg zeranol (Z) at intervals
     of 103 - 115 days from 12 weeks of age until 9 weeks
     before slaughter.
(c) Repeatedly implanted with 20 mg oestradiol plus 140 mg
     trenbolone acetate (O+TBA) as in (b)
(d) Repeatedly implanted with 20 mg oestradiol plus 200 mg
     progesterone (O+P) as in (b)
(e) Implanted with 45 mg oestradiol - 17B in silicone rubber
     (C-0) at 12 weeks of age.

The animals were housed in a slatted floor building in lots
of 25-30 and fed silage and concentrates throughout the
experimental period. There were approximately equal
numbers of animals from each treatment in each pen. The
animals were weighed at 6 - 8 week intervals and 98 animals
were slaughtered. Carcass and testicular weights were
obtained. To attempt to assess treatment effects on
behaviour, animals were assigned a riding score depending
on the amount of wear on the inside of the front legs,

which varied from zero for no wear to 2 for hair well worn
and skin exposed.

Variables were analysed using a least-squares regression
programme. Treatments were compared with covariate adjustments for initial weights. Carcass weights and testicular
parameters were analysed using days to slaughter as an
additional covariate.

RESULTS

Trial 1. Animals treated with C-O, Z, C-O + TBA and Z +
TBA had higher (p<0.01) daily liveweight gains than control
animals (Table 1). Animals treated with Z had higher
(P < 0.05) daily liveweight gains than animals treated with
C-O. An additive effect (p < 0.05) in daily liveweight was
obtained from animals treated with C-O + TBA and Z + TBA.

TABLE I   Initial weight, daily liveweight gain and carcass
weight of steers implanted once with oestradiol -
17B (C-O) or three times with zeranol (Z), Z plus
trenbolone acetate (TBA) or once with C-O and
three times with TBA.

TREATMENT	NO. OF ANIMALS	INITIAL WT. (kg)	DAILY LIVEWEIGHT GAIN (kg)	NO. SLAUGHT- ERED	CARCASS WEIGHT (kg)
Control	72	316	0.69^{1*}	57	300^{1}
C-O	80	320	0.75^{23}	57	306^{12}
Z	84	313	0.78^{3}	64	311^{23}
C-O + TBA	79	318	0.83^{4}	64	316^{34}
Z + TBA	83	310	0.86^{4}	67	321^{4}
Standard Error		5.0	0.02		

*Means with different superscripts are significantly
different (p < 0.05).

Pooled results for all animals treated with C-O or Z
irrespective of whether or not they were given TBA (±TBA)
are presented in Table 2. All animals treated with C-O ± TBA

had higher (p< 0.01) daily liveweight gains than control
animals and animals treated with Z ± TBA had higher
(p< 0.01) daily liveweight gains than animals treated with
C-O ± TBA.

TABLE 2.   Initial weight, daily liveweight gain and carcass
           weight of all animals implanted with oestradiol -
           17B (C-O) or zeranol (Z) with or without (±)
           trenbolone acetate (TBA).

TREATMENT	NO. OF ANIMALS	INITIAL WT. (kg)	DAILY LIVEWEIGHT GAIN (kg)	NO. SLAUGHT- ERED	CARCASS WEIGHT (kg)
Control	72	316	0.69^{1*}	57	300^1
C-O ± TBA	159	318	0.79^2	121	311^2
Z ± TBA	167	312	0.83^3	131	316^3
Standard Error		3.5	0.01		

*Means with different superscripts are significantly
different (p <0.01).

The carcass weights of animals treated with C-O were not significantly (p >0.05) heavier than carcass weights of control
animals (Table 1). The carcasses from animals treated with
Z and C-O + TBA were 11 kg and 16 kg heavier (p< 0.01) than
those control animals respectively. An additive effect
(p< 0.01) in carcass weight was obtained from the use of
repeated TBA in animals getting either C-O or Z. As regards
the pooled results (Table 2) all animals treated with C-O ±
TBA had significantly (p< 0.01) heavier carcasses than
control animals and animals treated with Z ± TBA had significantly (p< 0.01) heavier carcasses than animals treated with
C-O ± TBA. The conformation and fat scores of implanted
and control animals were not significantly (p> 0.05)
different.

     The ears of 36 C-O implanted animals were recovered and
of these 24 had implants present. In the remaining 12
animals where implants were not recovered, movement of
implant to a different position could not be eliminated.

The total liveweight gain and carcass weights for animals
from whom implants were recovered were 216 ± 4.8 kg and
306 ± 2.8 kg compared to 199 ± 3.4 kg and 294 ± 6.9 kg for
C-O implanted animals from whom implants were not recovered
after slaughter. These differences were not significent
(p > 0.05).

Trial 2 : The results are shown in Tables 3 and 4. Daily
liveweight gain was higher (p < 0.05) for animals treated
with Z, O+P and O+TBA compared to animals treated with C-O
and control animals (Table 3). Carcass weights from animals
treated with O+P and O+TBA were heavier (p < 0.05) than from
animals treated with Z and C-O and control animals.

TABLE 3   Initial weight, daily liveweight gain and carcass
           weight of bulls repeatedly* implanted with
           different anabolic agents.

TREATMENT	NO. OF ANIMALS	INITIAL WT. (kg)	DAILY LIVEWEIGHT GAIN (kg)	NO. SLAUGHT-ERED	CARCASS WT. (kg)
Control	28	122^1	0.71^1	22	208^1
Z*	27	124^1	0.79^2	18	215^1
O+TBA	27	120^1	0.81^2	18	224^2
O+P	26	122^1	0.79^2	21	225^2
C-O	25	122^1	0.74^1	19	209^1
SEM		3.5	0.02		5.0

*Zeranol (Z), oestradiol plus trenbolone acetate (O+TBA)
and oestradiol plus progesterone (O+P) were implanted three
times at a mean interval of 109 days. Oestradiol 17β
in silastic rubber (C-O) was implanted only at the start
of the experiment.
[1,2]Means with different superscripts are significantly
different (p < 0.05).

Animals implanted with Z, O+TBA and O+P had higher
(p < 0.05) liveweight gains after the first implant (Table 4)
Animals treated with C-O gained weight at the same rate as
control animals. Liveweight gain after the second implant

TABLE 4　　Effect of implantation of bulls on liveweight
　　　　　　gain (kg) during successive implant periods.

| TREATMENT | 1st IMPLANT | | 2nd IMPLANT | | 3rd IMPLANT | | TESTICULAR |
	N	103 DAYS	N	115 DAYS	N	74 DAYS	WT. (g)
Control	27	0.94^1	26	0.59^1	26	0.58^1	242^1
Z	27	1.06^2	27	0.63^1	27	0.64^1	212^2
O+TBA	27	1.02^2	26	0.70^2	26	0.66^1	194^2
O+P	26	1.04^2	26	0.66^2	27	0.65^1	211^2
C-O	25	0.94^1	25	0.64^1	25	0.58^1	227^1
SEM		0.04		0.035		0.05	15.0

*Zeranol (Z), oestradiol plus trenbolone acetate (O+TBA)
and oestradiol plus progesterone (O+P) were implanted three
times at a mean interval of 109 days.　Oestradiol 17β
in silastic rubber (C-O) was implanted only at the start
of the experiment.
[1,2]Means with different superscripts are significantly
different ($p < 0.05$).

was higher ($p < 0.05$) for animals treated with O+TBA and O+P
than for animals treated with Z, C-O and control animals.
There was no difference ($p > 0.05$) in liveweight gain
between treatments after the third implant.　Testicular
weight was greater for control animals and animals treated
with C-O than for animals repeatedly implanted with Z,
O+P and O+TBA.

　　　Animals on the different implant treatments had lower
($p < 0.05$) riding scores on day 265 than control animals.

DISCUSSION
　　　The results of trial 1 demonstrate that the C-O implant
is effective in yearling steers for at least 217 days.　There
was no significant difference in growth rate for C-O and
controls in the last 63 days of the experiment.　This
diminution in response could be due to failure of C-O

implants to continue to release sufficient oestradiol, or
to the loss rate of C-O implants. The large C-O implant
can be lost shortly after insertion and the tendency for
a difference in total liveweight gain and carcass weight
of C-O implanted steers who had or had not implants
recovered after slaughter, although only different at 10%
significance level, points to loss rate as a possible
contribution to reduced effectiveness of C-O implants.
The significantly increased growth rate from repeated
implantation with zeranol compared to a single implant of
C-O, with or without trenbolone acetate, may also be
due to loss rate of C-O implants rather than physiological
differences in effectiveness between Z and C-O implants.

The results demonstrate an additive effect on growth
of C-O implants in conjunction with the synthetic androgen,
trenbolone acetate. Repeated implantation with zeranol is
effective in increasing growth rate and carcass weight in
steers (Roche and others, 1981). Repeated implantation
with zeranol and trenbolone acetate together resulted in a
further significant increase in growth rate and carcass
weight compared with that obtained from repeated implantation
with zeranol alone. This confirms work by other authors that
additive effects are obtained from implantation of steers
with oestrogenic and androgenic compounds together (Heitzman
et al, 1977; Roche et al, 1978) and that these additive
effects are maintained with repeated implantations (Roche
and Davis 1983).

The effect of oestrogenic anabolic agents in bulls was
not as clear cut compared to responses obtained in steers.
A significant increase in daily gain was obtained. In
addition to the effects on growth, testicular weight and
development were reduced by all anabolic agents. This
reduction in testicular development was presumably responsible
for the reduced riding score, suggesting reduced mounting
activity in bulls implanted repeatedly with oestrogens.
This presumably reflects reduced gonadotrophic secretion
which in turn is responsible for lower levels of testosterone.

Therefore, it appears possible to castrate bulls hormonally
by repeated administration of oestrogens and obtain better
growth rate, reduced testicular development and reduced
mounting behaviour. This approach may be an alternative to
castration by physical means. The implications of reduced
mounting behaviour of implanted bulls on pre slaughter
stress may lead to a reduction in dark cutting meat in
slaughtered bulls which is caused by increased stress prior
to slaughter. It also means that in countries using bulls
for beef production consideration could be given to the use
of oestradiol in order to increase growth rate and ease
management problems by reducing agressiveness and mounting
behaviour, with possible benefits on meat quality.

REFERENCES

Convey, E.M., Beck, T.W., Neitzel, R.R., Bostwick, E.F. and
        Hafs, H.D. (1977). Negative feedback control of
        bovine serum luteinizing hormone (LH) concentra-
        tion from completion of the preovulatory surge
        until resumption of luteal function. J. Anim.
        Sci. 45 : 792-796.

Galbraith, H. and Topps, T.H. (1981). Effect of hormones
        on growth and body composition of animals. Nutr.
        Abstr. Rev. Ser. B51 : 521-540.

Heitzman, R.J., Chan, K.H. and Hart, I.C. (1977). Live-
        weight gains, blood levels of metabolites,
        proteins and hormones following implantation of
        anabolic agents in steers. Brit Vet. J. 133 :
        62-70.

Mies, W.L., Sherrod, L.B. and Elliston, N.G. (1979).
        Evaluation of a new removable implant for
        finishing steers. Proc. West. Sert. Anim. Sci.
        30 : 275-278.

Roche, J.F., Davis, W.D. and Sherington, J. (1978). Effect
        of trenbolone acetate or resorcylic acid lactone
        alone or combined on daily liveweight and carcass

weight in steers.   Irish J. Agric. Res. 17 :
7-14.

Roche, J.F., Harte, F.J., Joseph, R.L. and Davis, W.D. (1981).
The use of growth promoters in beef production.
Proc. of CEC Meeting 'Anabolic Agents in Beef
and Veal Production' Brussels, March 1981, p.
27-44.

Roche, J.F. and Davis, W.D. (1983).   Effect of re-implanting
anabolic agents on liveweight and carcass weight
of beef cattle.   Vet. Rec. 112 : 79-81.

DISCUSSION ON DR. ROCHE'S PAPER

Dr. Heitzman: In your data you got 21kg. more carcass
weight with repeat implantation of zeranol
and trenbolone acetate compared to 6kg. with Compudose alone.
On economics grounds is this a real difference
taking into account handling problems and cost of
implants.

Dr. Roche: To get maximum growth responses in steers, cattle
should be given a male and a female type anabolic agent.
Therefore the handling problems do not enter into the
discussion as Compudose implanted steers also
need to be given trenbolone acetate at 3 - 4 month intervals.
Hence, it is easy to give zeranol at the same time. From an
economic view point feed efficiency needs to be considered.
Loss rate of Compudose and behavioural problems also need
to be considered. We did get slightly better responses from
repeated zeranol than from Compudose alone with or without
trenbolone but loss of Compudose could explain
this.

Dr. Heitzman: We are interested in these data in the U.K.
since hexoestrol is now banned we are left with a choice
between zeranol and Compudose when advising producers.

Dr. Roche: In Ireland we have a further choice namely
20mg. oestradiol and 200mg. progesterone. We get better
growth rates with oestradiol-progesterone after a second
implant given to steers in July at pasture compared with
Compudose.

Dr. Forbes: In your bull trial, oestradiol + progesterone
did not affect daily gain but had a marked effect on
carcass weight. Was this due to gut fill effects?

Dr. Roche: No. There was a significant increase in daily gain after the second implant and an increase after the third and fourth implants which did not reach significance. It is this increased gain that gave the extra carcass weight. Killing out percent was not measured as the bulls were not weighed the day of slaughter and this factor must also be taken into account. On the second trial site where data were not shown we did get a significant effect of oestradiol-progesterone on dialy gain.

Dr. Quirke: Any behavioural problems following the use of Compudose or problems with loss of implants?

Dr. Roche: Loss of implants can be a problem with Compudose. Where we have palpated the ears and collected ears after slaughter losses can be as high as 5 - 20%. As regards behaviour, any implant containing oestradiol whether or not it is a silastic implant, is prone to give transient behavioural problems. In the case of the silastic implant, there are crystals of oestradiol on and very close to the surface. This initial burst of oestradiol does elevate blood levels for 2 - 3 days thus increasing the risk of behavioural problems.

Dr. Buttery: Do you have any evidence that Compudose was running out of oestradiol or is it possible that there was down regulation of oestradiol receptors?

Dr. Roche: I do not know whether or not there was down regulation. If this is a problem then why did it not occur after 90 - 100 days when Compudose was giving a significant increase in daily gain. We do not have evidence on blood levels of oestradiol 200 - 300 days after implantation but from the growth data, Compudose gave significant increase in daily gain from April to November. Thereafter, gain was not significant with Compduose alone or gain with Compudose and trenbolone was not better than zeranol alone. This suggests the possibility that Compudose is not releasing sufficient oestradiol after about 220 days and this point requires clarification. It may also be related to loss of implants.

Dr. Galbraith: We have investigated the growth response of steers to Compudose-365 and observed significant improvements in liveweight gain which were less than those obtained in the presence of Compudose-365 + trenbolone acetate, (300mg. Finaplix). We also collected blood samples from the jugular vein ipsilateral to the site of implant and the concentrations of steroids are shown in Fig.1.

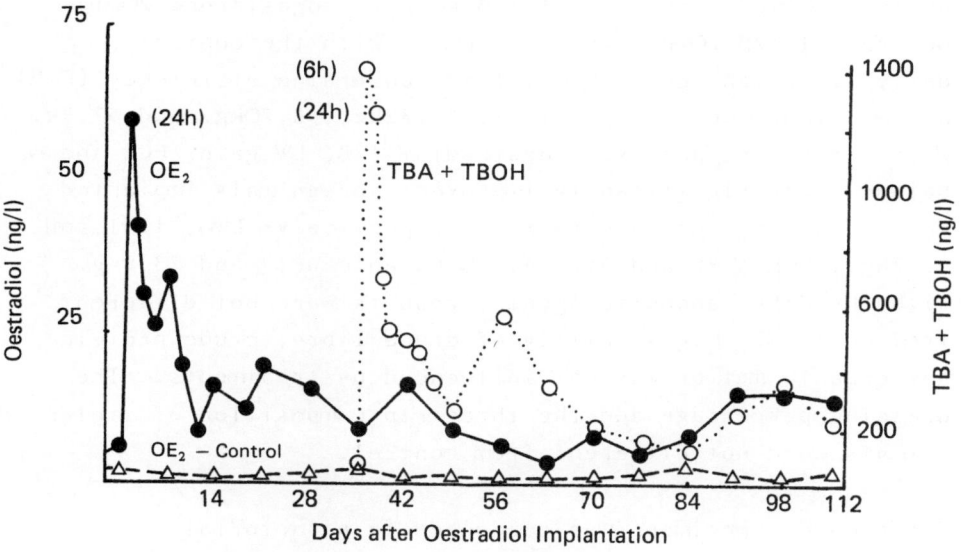

Fig. 1: Changes in the concentrations of oestradiol - 17β (OE_2) and trenbolone acetate (TBA) + trenbolone (TBOH) in steers (n = 4) implanted with Compudose-365 (45mg OE_2) at day 0 and Finaplix (300mg TBA) on day 35 of a 112 day experiment.

The concentrations of both oestradiol-17β and trenbolone peak within 24h of implantation and then appeared to stabilise within two to three weeks. Concentrations in the contralateral vein were lower than those in the ipsilateral vein throughout.

Dr. Bienfait: In two experiments, the effects of anabolic agents were studied on growing fattening young bulls over periods of 112 days. Animals were implanted once at 56 days or twice at 112 and 56 days before slaughter. A dry sugar beet pulp based ration was fed during the two experiments at the rate of 100g/kg metabolic live weight.

In the first experiment, anabolic agents were: trenbolone acetate 200mg. + oestradiol 17β 40mg., progesterone 200mg. + oestradiol 17β 20mg., zeranol 36mg. With the control, daily liveweight gain (LWG), food conversion efficiency (FCE) and N balance were respectively: 1.32kg., 6.70kg. and 57.9g. With trenbolone acetate + oestradiol 17β, LW gain, FCE and N balance were significantly improved; for animals implanted once or twice, the results were respectively: LWG, 1.51 and 1.67kg., FCE 5.78 and 5.72kg., N balance 66.9 and 73.9 g. With the other anabolic agents, results were not different from control. Digestibility of crude fibre, crude protein and organic matter was not influenced by treatments. The dressing percentage and the three rib composition of implanted animals were not different from control.

The second experiment was conducted in a factorial design 3 x 2 with trenbolone acetate 200mg. + oestradiol 17β 40mg., trenbolone acetate 200mg. + zeranol 36mg., testosterone 200mg. + oestradiol 17β 20mg. and two periods of implantation as before. No significant differences were observed between the treatments but the best results were obtained with trenbolone acetate + oestradiol 17β: LWG 1.50kg., FCE 6.48kg., N balance 65.5g. For the testosterone + oestradiol 17B and trenbolone + zeranol treatments, LWG, FCE and N

balance were respectively: 14.5kg., 6.62kg., 60.9g and
1.37kg., 7.05kg and 62.1g. With trenbolone + oestradiol, the
results were improved by the 112 days implantation.
With the other anabolic agents, no significant effects
of the repeated implantation were observed. In this
experiment, the urinary excretion of 3-Methyl histidine
was measured as an index of muscle-protein breakdown in
cattle. The results combined with N balance indicated that
the muscle-protein synthesis was greatly improved with
anabolic agents.

THE EFFECTS OF ANABOLIC AGENTS ON THE FIBERS OF THE
L. dorsi MUSCLE OF MALE CATTLE

M.J. Clancy, Janet M. Lester and J.F. Roche[*]

The Agricultural Institute, Animal Production Research
Centre, Dunsinea, Castleknock, Co. Dublin
and Grange, Dunsany, Co. Meath, Ireland

Summary

Samples of longissimus dorsi muscle were taken from carcasses
of steers, steers implanted with anabolic agents and from bulls,
of Friesian and Charolais cross Friesian breeds of cattle.
Percent and mean cross-sectional areas (CSA) of three myofiber
types (βR, αR, and αW) were determined.

The percentage of βR myofibers did not vary significantly with
treatment. The implanted steers had 26% more αR and 8% less αW
myofibers than the untreated steers, while the bulls had 33%
more αR and 20% less αW myofibers than the implanted steers
(P < 0.001).

In the implanted steers the mean CSA of the βR myofibers was
significantly greater than that of the untreated steers, but
did not differ from that of the bull. The mean CSA of the αR
myofibers increased considerably with treatment, but only that
of the bull was significantly greater than that of the untreated
steer. The mean CSA of the αW myofibers in the implanted steer
was identical with that of the untreated steer and significantly
smaller than that of the bull. In comparison to the untreated
steers, significant hypertrophy of all three myofiber types

[*] Present address: Veterinary Field Station, University College
Dublin, Ballycoolin Road, Finglas, Dublin, 11, Ireland.

occurred in bulls.

These findings demonstrate a significant increase in the
oxidative capacity of the longissimus muscle when the levels
of both endogenous and exogenous anabolic agents are increased.
They are also consistent with the greater efficiency of
deposition of protein obtained, with implanted steers and bulls.

No evidence of myofiber abnormality was found in any of the
samples examined.

Introduction

Anabolic agents are used in meat production to increase growth
rate, feed conversion to protein and the lean meat content of
carcasses (Galbraith and Topps, 1981). Thus, while the value
of anabolics for meat production is well established, rather
little is known about the effects these substances have on the
muscle fibers of meat animals. Indeed, basic factors which
limit the rate and extent of muscle growth and protein synthesis
as well as those concerned with adipocytes and fat deposition
in meat animals have not been defined. An understanding of
these factors in cellular and molecular terms would permit a
greater manipulation by scientists to optimise meat production.

  Studies on the effects of anabolic agents on the muscle fiber
types of meat animals are few. Fox et al. (1973) found that
the ratio of oxidative (red) to anaerobic (white) myofibers in
the longissimus dorsi muscle of Charolais-Hereford steers was
not significantly affected by Tapazole (1-methyl-2-mercapto-
imidazole), a feed additive which increases weight gains and
feed efficiency in beef. Spender et al.(1980) observed a sex
effect when they found that the proportion of oxidative fibers
in the biceps femoris muscle of heifers was significantly
greater than in the same muscle from steers. In both of these
studies, however, a non-specific oxidoreductase reaction
(NADA : tetrazolium oxidoreductase) was used to identify oxidative myofibers. There is, therefore, a need for a more detailed

study of the effects which anabolic agents may have on the
muscle fibers of meat animals. Accordingly, this paper describes
the effects endogenous and exogenous anabolic agents have on
the muscle fibers of the longissimus dorsi muscle from male
cattle.

A wide range of histochemical reactions were used to positively
identify three myofiber types :- β-Red (βR), α-Red (αR) and
α-White (αW), using the nomenclature of Ashmore and Doerr (1971).

Materials and Methods

Test animals

Sixteen of 24 Friesian and 16 of 24   Charolais cross Friesian
bull calves were castrated at 5 to 6 months of age. Animals
of both genotypes were allocated equally to the following
treatments :-

1)   Steers
2)   Steers implanted with anabolic agents in the ear, and
3)   Bulls

The steers were implanted with both 36 mg resorcylic acid
lactone and 300 mg trenbolone acetate, when they were turned
out to grass at 16 months of age. At the beginning of the
subsequent winter fattening period, these same steers were
re-implanted with 20 mg oestradiol benzoate and 200 mg progesterone
and with 300 mg trenbolone acetate.

All animals were grazed together at pasture from May to
November. In November each group of eight animals was housed
separately in a slatted floor house. They were fed grass
silage and concentrates until slaughtered at 26 months of age.
Live weights and carcass weights were recorded for each animal.

Sample collection

Half of the total number of carcasses were selected for sampling
in the factory. This gave a 3(treatments) x 2(genotypes)
x 4 animals factorial experiment. A steak (2.5 cm thick) was
removed from the position posterior to the last rib on the right

hand side of each carcass 24 h post mortem. The steaks were
taken to the laboratory where they were stored at 2°C for 24 h.
Pieces (1x1x2 cm) with parallel fasciculi were then cut at the
same medial site in each steak, frozen and stored in liquid
nitrogen for examination later.

pH measurement

The pH was measured on each steak at 24 h post mortem by inserting
a combination electrode into the tissue.

Myofiber histochemistry and histology

Myofibers were classified as β-Red (βR), α-Red (αR), or α-White
(αW) on the basis of their staining reactions for acid (pH 4.35)
and alkaline (pH 10.4) stable myosin Ca^{2+}- activated ATPases
(EC 3.6.1.3); succinate dehydrogenase (EC 1.3.99.1) (SDH) and
mitochondrial glycerol-3-phosphate dehydrogenase (glycerol-3-
phosphate : menadione oxidoreductase, EC 1.1.99.5) (GPOX) in
serial transverse cryosections (10 μm), which were cut in a
cryostat maintained at -15°C. The cryosections, mounted in
coverslips, were air-dried at room temperature for approximately
30 min before staining. The identification of the three myofiber
types was greatly facilitated by combining the SDH reaction with
(a) the acid-stable myosin ATPase on one cryosection and with
(b) the alkali-stable myosin ATPase on a second and by correlations
with the GPOX stain (Clancy and Lester, 1982).

Serial cryosections were also examined for glycogen by the
periodic acid-Schift method. For morphological examination,
cryosections were stained with Harris's haematoxylin and eosin;
Oil Red O combined with Mayer's haematoxylin and Gomori trichrome;
and Weigert's haematoxylin and picric ponceau.

Morphometry

Enlarged (x 250) images of the PAS-stained cryosections were
projected on to white paper using a projection microscope.
With the fasciculus as the sample unit the perimeter of each
myofiber cross-section was traced. The three myofiber types
(βR, αR and αW) were identified and enumerated, and their cross-
sectional areas were determined using a Zeiss particle size

analyser (Clancy and Herlihy, 1978). Six fasiculi were sampled
as outlined, amounting to between 200-300 myofibers of each type
per tissue sample.

Statistical analyses

Analysis of variance was used to determine the effects of
treatments and breed. The significance of differences between
means was determined by the stutentized range test (Snedecor,
1967). For the purposes of this paper, only the treatment
effects are reported here.

Results and Discussion

Growth and carcass parameters

Treatment means for animal growth and carcass characteristics
are given in Table 1. The implants, as expected, significantly
increased growth rate in steers, while the implanted steers did
not differ significantly in growth rate from bulls. The carcass
weight of the implanted steers differed significantly from that
of the untreated steers, but not from that of bulls.

The pH of the muscle at 24 h was the same for all three
treatments. This in conjunction with the visual appearance of
the tissue indicates that the post mortem glycolysis was normal
in all cases. Similar patterns of post mortem depletion of
glycogen in the myofibers was observed in all three treatments.

Myofiber morphology

No morphological irregularities were observed in any of the
histologically stained sections. The post mortem progression
from muscle to meat was normal in all cases and the cellular
structures of the meats were also normal. In particular, no
displaced nuclei, spiked or ragged myofibers were seen, which
are indicative of myopathies. Furthermore, the acid-stable
myosin ATPase reaction was in all cases the exact reversal of
the alkaline-stable form. The myofibers can, therefore, be
considered normal (Dubowitz and Brook, 1973).

Histochemical classification

The myofibers were classified according to their histochemical
staining reactions into three types :-

1) β-Red (βR) myofibers which possess,

    (i)   an acid-stable, alkali-labile myosine ATPase,

    (ii) a high oxidative capacity, as indicated by SDH, and

    (iii) a low to moderate activity as indicated by GPOX.

2) α-Red (αR) myofibers which possess,

    (i)   an acid-labile, alkali-stable myosin ATPase,

    (ii) with moderate to high oxidative and anaerobic activities,
       and,

3) α-White (αW) myofibers possessing,

    (i)   an acid-labile, alkali-stable myosin ATPase, and

    (ii) with low oxidative and high anaerobic capabilities.

The terms fast-twitch and slow-twitch (Peter et al., 1972) are
often used synonomously with alpha (α) and beta (β), but since
the physiological speeds of contraction of meat animal myofibers
have not been measured, the use of the terms α and β are preferable.

Changes in myofiber types

The percentages of βR myofibers and the percentage of alpha
fibres (αR + αW) remained constant when the level of both
exogenous (implanted steer) and endogenous (bull) anabolics
varied. Within the alpha group, however, there were significant
changes. In the progression from steers to implanted steers and
bulls, the percentage of the αR type was increased at the expense
of the αW fibers. This effect is seen by comparing steers, and
implanted steers and bulls with each other (Table 2). The
relationship between the decrease in the αW myofibers and the
increase in the αR myofibers is significant (P < 0.001) over the
three treatments and is given by :-

$$\% \ \alpha W \ = \ 74.0 - 1.0 \ (\% \ \alpha R) \quad s_{y.x} \ = \ 4.33$$

Since it is generally accepted that the real numbers of
myofiber in the muscles of meat animals are fixed at birth
(Swatland, 1976) it would appear that the increase in the
percentage of αR myofibers is due to a conversion of αW myofibers
to αR myofibers. While this work thus establishes the significant effects of anabolics on muscle metabolism, it remains
for further study to elucidate how this occurs.

Changes in myofiber size

Parallel with the changes in the percentages of the three
myofiber types, the sizes of the myofibers were affected (Table 3).
The mean CSA of the βR myofibers of implanted steers and bulls
was significantly greater than that of the untreated steers.
While there was also considerable hypertrophy of the αR myofibers
with treatment, only the mean CSA of αR myofibers from the bulls
was significantly greater than that of the untreated steers.
There was also a 20% increase (NS) in the size of the αR myofiber
of the implanted steer in comparison with that from the untreated
steer. Endogenous anabolic agents had no effect on the size
of the αW myofibers, yet these fibers were significantly
enlarged in the bull. It would appear, therefore, that the
relatively higher levels of testosterone, which occur in bulls,
are responsible for the hypertrophy of the αW myofibers, and
that exogenous agents fail to mimic this effect. This difference
between endogenous and exogenous anabolic agents in the bovine
requires further study.

By combining for each animal, the percentage of the different
fibers with their respective mean CSA, the proportions of total
area of muscle occupied by the three fiber types are obtained
(Table 4). These values give a measure of the changes in the
metabolism of the longissimus dorsi muscle as caused by anabolic
agents. Table 4 shows that the percentage area occupied by the
βR fibers, although increased, is not significantly affected by
treatment. However, the proportion of the area occupied by the
αR myofibers increases significantly and that of the αW decreases
($P < 0.001$) as the levels of anabolic agents increases. These
changes indicate that the oxidative capacity of the LD muscle
increases significantly when steers are treated with anabolic

agents, as used in this study, and that the oxidative capacity of bulls' longissimus muscle is greater still.

The energy requirement for protein synthesis, 1 or 2 ATP equivalents per amino acid residue for membrane transport, and 4 ATP equivalents per peptide bond, is very high and requires active energy metabolism (Trenkle, 1979). Since oxidative metabolism is 12 times as efficient as anaerobic metabolism in producing ATP from glucose residues, the significantly enhanced oxidative capacity of the LD muscle brought about by anabolic agents, can result in a more efficient production of ATP. This, in turn, can facilitate protein synthesis, which was manifest in the present work in the hypertrophy of the Red, or oxidative myofibers (Table 3).

This study, which appears to be the first to show that endogenous and exogenous anabolic agents significantly affect the proportions and sizes of muscle fibers in a meat animal, suggests one way in which anabolic agents result in a more efficient protein production in muscle, i.e. by increasing the efficiency of ATP productivity.

References

Ashmore, C.R. and Doerr, L. 1971. Comparative aspects of
    muscle fibre types in different species. Exptl. Neurol.
    31 : 408-418.

Clancy, M.J. and Herlihy, P.D. 1978. "Assessment of changes in
    myofiber size in muscle". In CEC Seminar on Patterns of
    Growth and Development in Cattle, Ghent, 1977.
    Ed. J. de Boer and J. Martin. Martinus Nijhoff, The Hague/
    Boston/London, pp 203-218.

Clancy, M.J. and Lester, Janet M. 1982. An approach to establi-
    shing the percentages of histochemical fiber types in
    skeletal muscle. J. Muscle Res. and Cell Motility, 3 :
    115-116.

Dubowitz, V. and Brook, M.H. 1973. Muscle Biopsy : A Modern
    Approach. W.B. Saunders & Co. Ltd., London, Philadelphia,
    and Toronto.

Fox, J.D., Moody, W.G., Boling, J.A., Bradley, N.W. and Kemp,J.D.
    1973. Effects of 1-methyl-2-mercaptoimidazole (Tapazole)
    feeding on muscle characteristics, fiber type and fatty
    acid composition of Charolais-Hereford steers.
    J. Anim. Sci. 37 : 438-442.

Galbraith, H. and Topps, J.H. 1981. Effect of hormones on the
    growth and body composition of animals. Nutr. Abstr.
    Rev. Ser. B51 : 521-540.

Peter, J.B., Barnard, R.J., Edgerton, V.R., Gillespie, C.A. and
    Stampel, G.E. 1972. Metabolic profiles of three fiber
    types of skeletal muscle in guinea pigs and rabbits.
    Biochemistry, 11 : 2627-2633.

Snedecor, G.W. 1967. Statistical Methods. 6th Ed. Ames, Iowa,
    U.S.A. The Iowa State Press.

Spinder, A.A., Mathias, M.M. and Cramer, D.D. 1980. Growth
    changes in bovine muscle fiber types as influenced by
    breed and sex. J. Fd. Sci. 45 : 29-31.

Swatland, H.J. 1976. Recent research on postnatal muscle
    development in swine. Proc. Rec. Meat Conf. 29 : 86-104.

Trenkle, A.H. 1979. Endocrinology of integrated lipid-protein
    metabolism in ruminants. Proc. Rec. Meat Conf. 32 : 87-92.

TABLE 1   Animal growth and carcass characteristics

	Treatments				
	Steer	Implanted steer	Bull	F-test	SE for treatment means
Live wt (kg) at 16 mts	333[a]	330[a]	359[b]	*	± 7.7
Slaughter wt (kg)	592[a]	636[b]	656[b]	***	± 9.3
Carcass wt (kg)	323[a]	347[b]	365[b]	***	± 5.9
Killing-out percentage	54.4[a]	54.6[a]	55.6[b]	*	± 0.32
Muscle pH at 48 h post mortem	5.53[a]	5.55[a]	5.57[a]	NS	± 0.03

Figures within a row followed by different superscripts are different at P < 0.05

TABLE 2   Percentages of muscle fiber types

Myofiber type	Treatments				
	Steer	Implanted steer	Bull	F-test	SE for treatment means
βR	25.9[a]	25.5[a]	26.9[a]	NS	± 1.58
αR	20.9[a]	26.3[b]	34.9[c]	***	± 1.67
αW	53.3[a]	48.0[b]	38.3[c]	***	± 1.80
(αR + αW)	74.1[a]	74.6[a]	73.2[a]	NS	± 1.58
(βR + αR)	46.8[a]	52.0[b]	61.8[c]	***	± 2.00

Figures within a row followed by different superscripts are different at P < 0.05

TABLE 3   Myofiber cross-sectional areas (μm sq)

Myofiber type	Steer	Implanted steer	Bull	F-test	SE for treatment means
βR	2250^a	2928^b	3284^b	***	± 153.6
αR	3211^a	3852^{ab}	4401^{bc}	*	± 269.3
αW	4729^a	4751^a	5699^b	*	± 252.4

Figures within a row followed by different superscripts are significantly different at P < 0.05

TABLE 4   Proportion by area occupied by myofibers[1]

Myofiber type	Steer	Implanted steer	Bull	F-test	SE for treatment means
βR	15.4^a	18.8^a	19.3^a	NS	± 1.42
αR	18.0^a	24.9^b	33.4^c	***	± 1.92
αW	66.6^a	56.3^b	47.3^c	***	± 2.00
(βR + αR)	33.4^a	43.7^b	52.8^c	***	± 2.00

Figures within a row followed by different superscripts are different at P < 0.05

[1]As a percentage

DISCUSSION ON DR. CLANCY'S PAPER

Dr. Schanbacher: Are there other data to confirm the sex effects on muscle fibre types and size?

Dr. Clancy: This seems to be the first report on clear cut effects on muscle fibres. There is a paper in Journal of Animal Science (Fox et al, 1973) that suggests these effects.

Dr. Buttery: In rats we have found exactly the same shift in fibre types following trenbolone.

Dr. Clancy: There is published work in mice also.

CARCASS QUALITY OF VEAL CALVES GIVEN ANABOLIC AGENTS

E.J. van Weerden

ILOB, Research Institute
in Animal Nutrition and Toxicology
Haarweg 8, P.O. Box 9
6700 AA Wageningen, the Netherlands

ABSTRACT

It can be concluded that there is a general agreement in the published reports that the application of anabolic agents, -mostly combinations of estradiol with testosterone or trenbolone-, has no distinct positive or negative effect on the carcass quality criteria meat colour, pH, dry matter, protein, collagen and fat contents of the meat; water holding capacity and loss after cooking or frying the meat. The conformation or meatiness of the carcasses is in general improved after anabolic treatment. For the tenderness of the meat, estimated by physical means via shear force and by taste panel evaluation, no systematic trends can be established at this stage. Whereas the tenderness of the M.long.dorsi seems to be affected slightly unfavourably by the anabolic treatment of the calf, two other muscles did not show any negative effect. So it is doubtful whether real differences exist in the quality of the meat from anabolic treated and untreated calves and, if they exist, the differences will be so small that it is unlikely that consumers of veal will detect them.

INTRODUCTION

A considerable number of literature data demonstrate the important improvement in the performance of cattle, especially steers, obtained by the application of anabolic agents (Galbraith and Topps, 1981). The positive effects of certain anabolic agents or combinations of anabolic agents on protein deposition, weight gain and feed conversion efficiency in veal calves are also well documented (van der Wal, 1975). With regard to the nature of the positive effect on live weight gain, experimental results of respiration trials in intact male veal calves, steers and castrated male pigs were discussed at the previous workshop on anabolic agents in Brussels (van Weerden et al., 1981). In all three categories of animals, the application of anabolic agents did not affect an increased total energy deposition per 1000 kcal intake of metabolisable energy, but resulted in an increased protein deposition and an, -energy equivalent-, decrease of fat deposition, so a shift from fat to protein deposition. As a consequence the total carcass of anabolic treated veal calves will contain more protein (meat) and less fat.

The question under discussion in the present paper is how these

phenomena of increased protein and decreased fat deposition in the carcass
are expressed in criteria determining slaughter and meat quality of the
veal calf. In the following presentation the criteria relating to the
quality of the slaughtered product were split up into two parts:
1) effects of anabolics on the total carcass; this means dressing percentage, colour of carcass, fatness and meatiness or conformation of the
   carcass.
2) effects of anabolics on the quality of the meat; this means chemical
   meat composition, physical parameters like waterbinding capacity,
   cooking loss, shear force, and the sensory assessment by taste panel.

EFFECTS OF ANABOLICS ON CARCASS QUALITY

Compared with the rather great number of literature data regarding the
effect of anabolic agents on performance, relatively few published reports
describe observations on carcass quality.

In an experiment of the group of Verbeke (1975) with veal calves of
approx. 180 - 190 kgs live weight at slaughter, an extensive examination of
the carcasses was carried out. The animals were treated 4 or 8 weeks before
slaughter with the combinations estradiol + testosterone or estradiol +
trenbolone acetate, part of the calves were not treated and were used as
negative controls. Carcasses from the treated animals were significantly
heavier than the control calves. Visual judgement of the meatiness showed
a significant improvement of the carcasses of the treated animals and also
the carcass index (ratio carcass weight : carcass length) was significantly
improved. Carcass colour was not affected by anabolic treatments.

Grandadam et al. (1975b)· and Scheid et al. (1974) have published
results in veal calves treated with 20 mg estradiol-17β + 140 mg trenbolone
acetate. Commercial carcass grading, which is mainly determined by meatiness, fat deposition and meat colour, were significantly better for the
treated calves compared with untreated control animals. This effect was for
the greater part caused by a more favourable meatiness of the carcasses,
attributed especially to a greater development of the posterior muscles.

The results of the judgement of carcass quality mentioned in the
literature, are in agreement with experimental data given in the following
two tables. In these tables all the results of experiments carried out at
ILOB and obtained during the last 10 years with the anabolic combinations
estradiol-17β + testosterone and estradiol-17β + trenbolone acetate are
compiled.

114

TABLE 1   Slaughter quality characteristics, I

n/treatment (total)	Dressing %		Colour (abs.)		Meatiness (abs.)	
	Control	Est.+test.	Control	Est.+test.	Control	Est.+ test.
11	66.4	65.8	7.5	7.4	6.9	6.7
12	67.1	67.7	8.1	8.0	7.0	7.7
15	66.5	66.9	7.7	8.0	6.8	7.1
8	65.8	65.5	6.6	7.1	6.3	7.1
11	66.4	65.8	7.5	7.4	6.9	6.7
10	66.5	67.3	8.0	7.8	7.2	7.4
67	66.4	66.5	7.6	7.6	6.8	7.1

In six experiments with 67 calves implanted with 20 mg
estradiol-17β + 200 mg testosterone, and 58 untreated control calves,
slaughter quality was determined by visual judgement by experts of the
colour and the meatiness of the carcasses (Table 1). For both criteria the
judgement was on a scale from 1 (very bad) to 10 (top quality). In addition
dressing percentage (warm carcass weight as a percentage of live weight
just before slaughter) was determined. Live weight at slaughter varied
between 160 and 200 kgs. Live weight at slaughter of the treated animals
was on average c. 3½ kg higher than that of the untreated controls. The
difference between treated and untreated animals is smaller than is normally found; this can be explained by the fact that, because of other
reasons, the calves were in most cases not slaughtered at the time when the
effect of the anabolic treatment was maximal.

It can be concluded that whereas no differences in dressing percentage
and colour were observed, meatiness was on average slightly better in the
treated animals.

In Table 2 the results of 11 experiments with the combination estra-
diol-17β (20 mg) + trenbolone acetate (140 mg) are compiled. Live weight
at slaughter varied between 160 and 200 kgs; the difference in live weight
between the groups at slaughter time was on average approx. 4 kgs, again a
relatively small difference because in most cases the calves were
slaughtered after the time of maximal effect. Again no difference in
dressing percentage and no systematic effect on carcass colour, but in all
11 experiments a better appraisal of the meatiness (conformation) of the
carcasses of the treated group was observed, differing from 0.1 to 1.1

units.

TABLE 2   Slaughter quality characteristics, II

n/treatment (total)	Dressing %		Colour (abs.)		Meatiness (abs.)	
	Control	Est.+trenb.	Control	Est.+trenb.	Control	Est.+trenb.
11	66.4	66.4	7.5	7.9	6.9	7.5
12	67.1	67.8	8.1	8.0	7.0	8.1
10	66.8	66.8	7.0	7.2	6.4	6.6
10	68.0	67.5	7.3	7.2	6.9	7.0
10	67.7	67.8	7.1	7.2	6.8	7.2
11	66.4	66.4	7.5	7.9	6.9	7.5
15	65.5	65.7	5.5	6.1	5.6	5.7
12	66.4	66.6	7.9	8.2	7.1	7.5
12	66.7	67.4	8.3	8.3	7.4	7.6
12	65.4	65.8	7.5	7.4	6.6	6.8
10	63.7	63.9	6.6	7.1	6.6	7.2
125	66.4	66.6	7.3	7.5	6.7	7.2

    Summarizing the results of the effects of anabolic agents on (external) carcass quality, it can be stated that there is   general agreement that whereas dressing percentage and carcass colour are not clearly affected, the meatiness or conformation of the carcasses of the treated calves is in most cases better. This conclusion is in agreement with the general opinion in practice, that anabolics especially improve the conformation of the carcass.

EFFECTS OF ANABOLICS ON MEAT QUALITY

    The number of published reports on the possible effects of the application of anabolic agents on the quality of calf meat is again very limited. There is a general agreement in the reports that the dry matter, N and ash content of the meat is not influenced; in two publications the fat content of the meat tended to be slightly lower (Verbeke et al., 1975; Schulz et al., 1976; Grandadam et al., 1975a; Valin et al., 1978). In one report no effect on the collagen (= connective tissue) content of the meat was observed, in a second report a slight tendency of an increased collagen content is mentioned and in a third report the collagen content of one muscle (Musculus longissimus dorsi) was higher after anabolic treatment, whereas in another muscle (Musculus pectoralis profundus) no difference could be detected

116

(Schulz et al., 1976; Verbeke et al., 1975; Valin et al., 1978).

Physical parameters of meat quality determined in the evaluation of the effect of anabolic agents on veal, are: pH, water holding capacity, weight loss at boiling or frying and shear force as an objective measure of tenderness of the meat. The published research reports agree that no distinct differences can be detected between control and anabolic treated calf meat samples in the criteria pH, water holding capacity, cooking loss and loss at frying (Verbeke et al., 1975; Fischer and Schröder, 1976; Valin et al., 1978; Grandadam et al., 1975a). For shear force, Verbeke et al (1975) found a tendency for higher values in the M. longissimus dorsi of treated calves. The french group at Beaumont (Valin et al., 1978) could not detect differences in two muscles of treated and non-treated calves in the maximum force values, the usual parameter measured, but in the M. longissimus dorsi the total quantity of work required to divide a sample of a standard core of meat was significantly greater for the anabolic treated meat; in het M. pectoralis profundus, however, no significant difference was observed.

In all the reviewed research reports on the subject of the assessment of meat quality after application of anabolic agents, the alarmingly high variability in the parameters studied becomes evident, a variability not only between animals of the same treatment group, but also between samples of meat of the same animal, so both a between-animal and a within-animal variability. This phenomenon makes it difficult to detect relatively small effects of treatment, reliable conclusions can only be drawn after a considerable number of observations has been made.

Results of sensorial assessment of meat quality after anabolic treatment are very scarce. Whereas not-officially published data of taste panel assessments, carried out by ITEB in France, showed no differences in appreciation between controls and anabolic treated veal, a study of Valin et al., (1978) indicates a lower appraisal of treated veal from the M. longissimus dorsi for tenderness and a tendency for a lower appraisal for juiciness and flavour.

To collect more data on the slaughter and meat quality of anabolic treated veal calves, in two experiments carried out at ILOB, examinations for these criteria were carried out.

In the first experiment, A, two control groups of 10 calves each and two test groups of 10 calves each were implanted with the combination estradiol + trenbolone acetate. Half of the calves were fed a ration with

16% crude protein and the other half a ration with 20% cp. The results for weight gain and feed conversion at 18 weeks of age, just before slaughter at c. 170 kgs live weight, are shown in Table 3.

TABLE 3   Weight gain and feed conversion, expt. A
(period 11-18 weeks of age, 7 weeks after implantation)

	Weight gain		Feed conversion	
	kg	%	abs.	%
Control, 16% cp	67.3	100	1.68	100
20 mg est.-17β + 140 mg trenb. ac., 16% cp	71.1	105.6	1.64	97.6
Control, 20% cp	69.4	100	1.65	100
20 mg est.-17β + 140 mg trenb. ac., 20% cp	74.3	107.1	1.60	97.0

For the examination of the meat, samples of the M. longissimus dorsi were collected from the left and right side between the 6th and 8th rib. All the analyses and determinations in the meat were carried out by CIVO-TNO, Department of Meat Technology.
Table 4 gives the analysis figures for dry matter and protein.

TABLE 4   Analysis of the meat (in%), expt. A

	Dry matter	Protein
Control, 16% cp	27.8	19.7
Est. + trenb., 16% cp	27.2	20.0
Control, 20% cp	28.3	19.5
Est. + trenb., 20% cp	27.9	20.4*

* significantly ($P < 0.05$) different from control.

In both treated groups dry matter content is slightly lower and protein content slightly higher, in one case the difference was significant.

As can be seen in Table 5, no appreciable differences were observed between controls and treated groups in water binding capacity, weight loss at boiling and shear force.

TABLE 5   Characteristics of the meat, expt. A

	Water binding cap. (%)	Weight loss at boiling (%)	Shear force values (in N*)
Control, 16% cp	31.1	25.1	33
Est. + trenb., 16% cp	31.6	25.5	35
Control, 20% cp	29.6	25.0	38
Est. + trenb., 20% cp	29.7	25.1	41

* shear force expressed in Newtons

In a second experiment (B) three groups of 10 calves each were in-volved. Table 6 gives the treatments and the results for weight gain and feed conversion at 20 weeks of age, just before slaughter at a live weight of c. 200 kg.

TABLE 6   Weight gain and feed conversion, expt. B
           (period 13-20 weeks of age, 7 weeks after implantation)

	Weight gain		Feed conversion	
	kg	%	abs.	%
Control	75.2	100	1.67	100
36 mg zeranol + 140 mg trenb. ac.	80.0	106.4	1.56*	93.4
20 mg est.-17β + 140 mg trenb. ac.	83.2*	110.6	1.49*	89.2

* significantly (P < 0.05) different from control

In Table 7 the analysed dry matter and fat contents in three muscles (M. longissimus dorsi, M. pectoralis profundus and M. rectus femoris) are given.
The M. pectoralis profundus has a higher fat content than both other muscles. No systematic differences related with treatment were observed.

Table 8 gives the figures for protein and collagen (oxyproline % x 8) content of the three muscles. Again no systematic effect of the treatment on the composition of the meat can be detected.

Shear force was measured in both raw and cooked samples of the three test muscles. Shear force is systematically higher in the M. pect. prof. than in both other muscles. The differences between the groups are always small and no systematic trends can be detected (Table 9).

TABLE 7   Analysis of the meat (in %), expt. B

	Dry matter	Fat
M. long. dorsi		
Control	24.4	1.6
Zer. + trenb.	24.4	1.3
Est. + trenb.	23.8	1.3
M. pect. prof.		
Control	24.0	2.7
Zer. + trenb.	24.8	3.1
Est. + trenb.	24.1	2.7
M. rect. fem.		
Control	23.7	1.6
Zer. + trenb.	23.5	1.4
Est. + trenb.	23.7	1.1

TABLE 8   Analysis of the meat (in %), expt. B

	Total protein	Collagen	Collagen free meat protein
M. long. dorsi			
Control	24.2	0.72	23.5
Zer. + trenb.	23.6	0.40	23.2
Est. + trenb.	23.6	0.56	23.0
M. pect. prof.			
Control	22.2	0.88	21.3
Zer. + trenb.	22.3	0.80	21.5
Est. + trenb.	22.7	0.96	21.7
M. rect. fem.			
Control	21.3	0.56	20.7
Zer. + trenb.	22.3	0.40	21.9
Est. + trenb.	22.3	0.88	21.4

TABLE 9   Shear force values (in N) of raw and cooked meat, expt. B
(mean values ± SD)

	Raw meat	Cooked meat
M. long. dorsi		
Control	20 ± 3	26 ± 6
Zer. + trenb.	26 ± 5	27 ± 6
Est. + trenb.	21 ± 3	28 ± 3
M. pect. prof.		
Control	50 ± 12	41 ± 6
Zer. + trenb.	48 ± 9	34 ± 9
Est. + trenb.	47 ± 12	36 ± 9
M. rect. fem.		
Control	28 ± 6	25 ± 3
Zer. + trenb.	25 ± 11	25 ± 6
Est. + trenb.	25 ± 8	27 ± 4

The sensory assessment of the tenderness of the meat was carried out
in cooked meat by a taste panel of eight trained persons. The panel members
were asked to rank the samples of the three test muscles from tender to
tough and thereafter to give a score to the samples in the scale from 1
(very tough) to 10 (very tender). The results (mean values ± SD) are compiled in Table 10.

TABLE 10   Mean scores for tenderness of cooked meat, expt. B
(1 = very tough, 10 = very tender)

	M. long. dorsi	M. pect. prof.	M. rect. fem.
Control	7.7 ± 0.55	5.8 ± 0.34	7.5 ± 0.74
Zer. + trenb.	7.4 ± 0.30	5.7 ± 0.39	7.1 ± 0.60
Est. + trenb.	7.2* ± 0.39	6.1 ± 0.52	7.4 ± 0.48

* significantly ($P < 0.05$) different from control.

The fact that the appreciation for tenderness of the M. pectoralis profundus
is systematically lower than the appreciation for the other two muscles is
in agreement with the higher shear force values of this muscle. For the M.
pectoralis profundus and the M. rectus femoris no distinct differences or
trends between the treatment groups can be seen, but for the M. longissimus

dorsi the appreciation of tenderness was significantly higher for the control group than for the estradiol + trenbolone treated group.

These results seem to indicate a difference in reaction of individual muscles to an anabolic treatment and also clearly illustrate the considerable variability within the animal for the criteria measured. Whereas our results with the M. longissimus dorsi agree with Valin's observations in this muscle, the finding that the tenderness of the other two test muscles in our experiment is not affected at all, throws doubt on the real significance of these observations for the appreciation of veal of treated calves as a whole.

REFERENCES

Fischer, A. und Schröder, K. 1976. Beeinflussung verschiedener Kriterien der Qualität von Kalbfleisch durch den Einsatz östrogenwirksamer Präparate in der Mast. Fortschritte in der Tierphysiologie und Tierernährung, Heft 6, 59-65.

Galbraith, H. and Topps, J.H. 1981. Effect of hormones on the growth and body composition of animals. Nutrition Abstracts and Reviews-Series B, 51, 8, 521-540.

Grandadam, J.A., Scheid, J.P., Dreux, H. et Deroy, R. 1975a. Influence de différentes préparations anabolisantes sur la qualité de la viande de veau. Rec. Méd. Vet. 151, 6, 355-362.

Grandadam, J.A., Scheid, J.P., Jobard, A., Dreux, H. and Boisson, J.M. 1975b. Results obtained with trenbolone acetate in conjunction with estradiol-17β in veal calves, feedlot bulls, lambs and pigs. Journal Animal Sci. 41, 969-977.

Scheid, J.P., Grandadam, J.A., Boisson, J.M. et Dreux, H. 1974. Expériment clinique d'une nouvelle préparation anabolisante dans la production du veau de boucherie. Rec. Méd. Vet. 150, 1, 37-48.

Schulz, V., Brandt, A. und Erbersdobler, H. 1976. Der Einflusz von Anabolika auf die Fleischzusammensetzung von Mastkälbern. Fortschritte in der Tierphysiologie und Tierernährung, Heft 6, 53-58.

Valin, C., Renerre, M., Touraille, C., Koop, J., Sornay, J. 1978. Influence de la forme d'apport de l'énergie dans la ration et de l'utilisation d'anabolisants sur la qualité de la viande de veau. Ann.Nutr. Alim. 32, 857-868.

Verbeke, R., Debackere, M., Hicquet, R., Lauwers, H., Pottie, G., Stevens, J., van Moer, D., van Hoof, J. and Vermeersch, G. 1975. Quality of the meat after the application of anabolic agents in young calves. In: Environmental Quality and Safety, Suppl. Vol.V, Eds. Georg Thieme, Stuttgart, 123-130.

Van Weerden, E.J., Berende, P.L.M. and Huisman, J. 1981. Application of endogenous and exogenous anabolic agents in veal calves. Proc. of an E.C. workshop in Brussels, March 5th and 6th, 1-26.

EFFECT OF IMMUNISATION AGAINST SOMATOSTATIN
ON GROWTH RATE OF LAMBS

G.S.G. Spencer

Animal Physiology Division
Agricultural Research Council, Meat Research
Institute, Langford, Bristol BS18 7DY U.K.

ABSTRACT

Immunisation of animals against somatostatin can produce marked effects on animal performance. This treatment has been found to be effective in certain breeds of sheep where quite considerable increases in growth have been observed. In a small study using St. Kilda lambs a 76% increase in rate of weight gain was observed in treated animals compared with controls. In a larger subsequent study on Dutch moor-sheep similar stimulation of weight gain was observed. The growth was not confined solely to weight gain; the standing height of the lambs was also significantly increased and food conversion efficiency was better in the treated lambs. Carcase dissection revealed that these changes occurred without any detrimental effects on body composition since muscle, fat and bone were all increased in proportion. It is still far from clear how this improvement in growth and efficiency is brought about; neither is it unequivocally proven that this treatment will be applicable to all breeds and species. However, the potential for growth stimulating effects is clear and it seems probable that, because of the wide range of effects of somatostatin, such things as milk production and wool growth may also be affected. The present experiments should be considered only as indicating a possible treatment for use in animal production; confirmation and extension of these studies is still required.

INTRODUCTION

For a long time pituitary somatotropin (GH) was considered to be the hormone that produced growth. It is now believed that the direct stimulators of growth at the cellular level are the growth hormone-dependent somatomedins (Daughaday et al, 1972). It is the importance of somatomedins as mediators of the actions of GH that explains,

in part, the remarkably poor stimulation of growth generally
obtained with exogenous GH administration to normal animals
(Machlin, 1976). Although the somatomedins are growth
hormone-dependent it seems likely that the correct balance of
hormones is required for maximal somatomedin production by
the liver (Wallis, 1980). There are many instances where GH
levels are high but, because of a general hormonal imbalance,
plasma somatomedin activity is decreased rather than
increased (Spencer et al, 1982). For example, in conditions
of severe malnutrition, growth hormone levels are elevated
but somatomedin levels are low (Grant et al, 1973).
Clearly, when nutrient availability is low the body needs to
divert energy away from growth while maintaining the
diabetogenic and lipolytic actions of GH to supply the vital
energy substrates required for maintenance; hence the
disparity between GH and somatomedin levels. Thus to
stimulate somatomedins production and growth a more
complicated manipulation of hormone levels is required.
Perhaps the easiest way to bring about such a change is to
manipulate the central control mechanisms which regulate
hormone release. Somatostatin may play a particularly
important part in this regard.

Somatostatin was originally identified as the
hypothalamic growth hormone-release inhibiting factor
(Krulich et al, 1972). It is now known, however, that it is
produced not only by the hypothalamus but also from the
pancreas and the gastro-intestinal tract (Arimura and
Fishback, 1981). The widespread production of somatostatin
is reflected in its diverse actions. It is now known to
affect not only GH release but also insulin (Koerker et al,
1974), glucagon (Ruch et al, 1973), thyroid stimulating
hormone (Vale et al, 1974), gastrin (Bloom et al, 1974) and
many other gastro-intestinal hormones (Schusdziarra, 1980).
It may, therefore, be particularly pertinent in growth
control - especially, perhaps, through GH, insulin and
thyroid hormones and thus increased somatomedin production.

Since somatostatin is an essentially inhibitory peptide,
it might be postulated that removal of this inhibitory

effect would lead to an increased growth rate. Such an approach has been undertaken using active immunisation against somatostatin. Such treatment has resulted in increased levels of hormones (Ferland et al, 1976 ; Varner, Davis and Reeves, 1980; Spencer and Williamson, 1981) and growth (Spencer and Williamson, 1981; Spencer, Garssen and Hart, 1983).

EXPERIMENTAL INVESTIGATIONS

In an initial experiment twin St. Kilda lambs were used. One lamb from each pair was immunised against somatostatin conjugated to human serum α-globulin while the other was immunised against globulin alone. The primary immunisations were made in Freunds Complete Adjuvant at

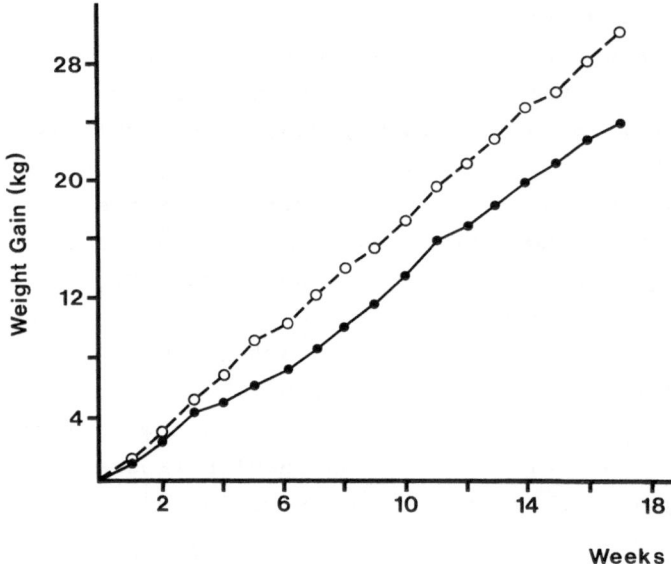

Fig. 1. Rate of weight gain (weight gain over initial weight) for lambs immunised against somatostatin (O----O) and those immunised against globulin (●——●).

three weeks of age and subsequent injections were in Freunds
Incomplete Adjuvant at fortnightly intervals
(Spencer and Williamson, 1981). The rate of growth of the
treated lambs (in terms of increased body weight) was
increased to 176% of that of the controls (8.27 ± 1.68 kg and
4.67 ± 1.36 kg weight increase over 9 weeks in treated and
controls, respectively).

A second experiment was undertaken at the Research
Institute for Animal Husbandry in the Netherlands. In this
experiment Dutch moor-sheep were used. A similar treatment
procedure and schedule was used (Spencer, Garssen and Hart,
1983) and, as in the first experiment, growth rate was
increased (Figure 1). It was found that it was not only body
weight which was increased; the stature of the lambs
(standing withers height) was also greater in the treated
than the control lambs (Figure 2).

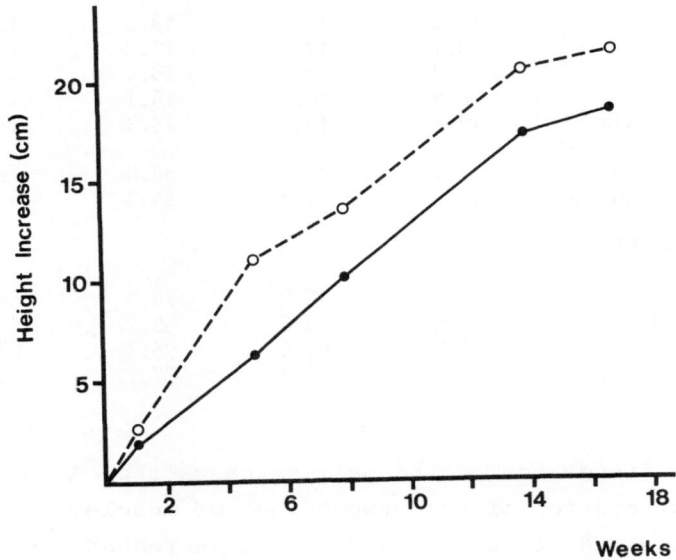

Fig 2. Increase in standing height (over initial standing
height) in lambs immunised against somatostatin (O----O)
and those immunised against globulin (•——•).

The lambs were killed at 20 weeks of age and the increased growth was confirmed by carcase dissection post-mortem. The dissection data revealed that the bones were bigger in the treated animals. For any particular bone it was difficult to show statistically significant increases in length because of the relatively small numbers in each group when separated according to sex and treatment. However, the increased sum of the length of complementary limb bones (femur and tibia/fibula, for example) clearly contributed to the increased stature seen in the treated animals (Table 1).

TABLE 1. The lengths and weights of the long bones from treated and control animals separated according to sex. Values are means. In both treated and control male groups n = 5; in both female groups n = 4; * p < 0.05.

| | Male | | Female | |
Bone length	Treated	Control	Treated	Control
Humerus	14.3	13.5	13.5	12.8 *
Ulna	18.1	17.4	17.2	16.6
Radius	13.9	13.4	13.1	12.7
Femur	16.9	16.4	16.1	15.6
Tibia/Fibula	19.3	18.4	18.3	17.2 *
Humerus & Ulna	32.4	29.4 *	30.8	29.3 *
Femur & Tib/Fib	36.2	34.8	34.4	32.8 *
Bone weight				
Humerus	84.8	75.8	70.3	58.0 *
Radius/Ulna	66.8	63.6	56.7	49.3 *
Femur	115.8	103.6	95.7	87.8
Tibia/Fibula	96.6	87.2	77.7	71.0

Complete dissection of the half-carcases showed that there was no alteration of the proportions of muscle, fat and bone in the treated lambs (Table 2). The increased growth had, therefore, been stimulated proportionally and was not due solely to excessive fat deposition.

Although the growth stimulating effect is of importance in itself, it is desirable that this is not bought at the expense of excessive food intake. Although this is not particularly crucial in pastured lambs it is a consideration for feed-lot cattle and for animals, such as pigs, fed high concentrate diets. The food intake data for both groups are shown in Table 3. From these data it is clear that the food intake/Kg body weight was decreased by 10% in the treated lambs (71.6 MJ/Kg compared with 78.6 MJ/Kg for the controls). Since the lambs were offered food ad libitum, this reflects a change in relative appetite (appetite related to body weight). Associated with this was an improvement in food utilisation (energy intake/change in body weight) in the treated lambs.

As a consequence of their increased growth rate, the treated lambs had reached 30 Kg (the slaughter weight of the control animals) by 16 weeks of age rather than the 20 weeks taken for the controls to reach this size (a 20% saving in time taken to slaughter). This has an effect on overall

TABLE 2. Proportions of muscle, fat and bone in carcasses of lambs immunised against somatostatin and control lambs immunised against globulin alone. All lambs were slaughtered at 20 weeks of age.

	CONTROL	TREATED
Carcass weight	14.42 ± 0.95	17.08 ± 0.82
Basic side weight	6738 ± 453	7931 ± 390
Muscle weight	3588 ± 209	4181 ± 266
Bone weight	980 ± 64	1093 ± 63
Fat weight	1987 ± 196	2453 ± 148
% muscle	54	53
% bone	15	14
% fat	29	31
Muscle : bone	3.68	3.82
Muscle : fat	1.90	1.73

efficiency. It is perhaps more informative to study food intake for the same weight gain in both groups. The increase in efficiency is even more pronounced when food intakes in the control and treated lambs are compared up to the same weight rather than age (Table 3). This finding raises the question of how the increased growth rate is brought about.

TABLE 3. Food intake and utilisation in control lambs immunised against globulin and in treated lambs to (a) the same age and (b) to the same weight.

	CONTROL (To Slaughter)	TREATED (To Slaughter)	TREATED (To 30 kg)
Change body weight	24.34	29.61	24.60
Mean food intake (g/lamb/day/kg wt)			
Milk	4.8	4.7	6.7
Pellets	31.0	27.4	25.1
Hay	5.7	5.7	5.2
Total energy intake (MJ/lamb)	1383.3	1470.3	1018.1
(MJ/lamb/kg wt)	78.6	71.6	56.5
Food utilisation (MJ / Δ wt)	56.83	49.66	41.39

It had been proposed that reduced inhibition of GH, insulin and thyrotropin release would lead to increased somatomedins and thus increased growth. However in these particular studies, in contrast to previous reports, no increase in basal GH, insulin or thyrotropin levels were observed - at least at fifteen weeks after treatment began. Somatomedin levels, however, were elevated in the treated lambs and there was an increased stimulation of growth hormone release following arginine infusion in the treated lambs (Spencer, Garssen and Hart, 1983). It is known that there are potent feedback mechanisms by which the body attempts to maintain endocrine balance. It is possible that the lambs had regained a degree of hormonal balance by fifteen weeks of treatment. Measurements of hormone levels earlier in the experiment might have revealed elevated plasma growth hormone levels and, indeed, the most striking increases in growth velocity were observed earlier in the treatment. Other workers, however, have found hormone changes this late in immunisation studies (Varner et al, 1980), and increased somatomedin levels were found in these lambs (Spencer, Garssen and Hart, 1983).

The way in which this immunisation procedure stimulates growth is, therefore, still to be elucidated. However, it is known that as well as inhibiting the release of pituitary and pancreatic hormones, somatostatin also has effects on the following gastro-intestinal tract functions: gastrin and gastric acid secretion, (Bloom et al 1974) motilin and gastric emptying (Bloom et al 1975), duodenal motility (Boden, Jacob & Staus, 1976) and the release of pepsin (Gomez-Pan et al 1975), secretin (Boden, Sivitz, Owen and Essa-Kouman, 1975), and cholecystokinin (Schlegel et al, 1977). It is possible, therefore, that the treatment had a major effect on removing inhibitory actions (Schusdziarra, 1980) of somatostatin on gut motility and nutrient absorption from the gastro-intestinal tract. Although this has not been directly investigated in detail in these lambs, it remains a possible effect which could explain the increased food utilisation efficiency and growth of these animals in the apparent absence of markedly elevated levels of growth hormones.

Both the St. Kilda and Dutch moor-sheep breeds are relatively unimproved breeds of sheep. From the practical point of view it is important that this treatment is effective in commercial breeds and also in other species. Somatostatin is not species specific and it would seem probable, therefore, that the technique could be equally well applied to cattle and pigs. A preliminary experiment (in collaboration with Dr. Diane Williamson) investigating both these points has been carried out in Large White pigs.

130

Although only a small pilot investigation, it was clear that
the somatostatin-immunised pigs grew faster than the control
animals (Figure 3). Unfortunately, an attempt to repeat this
experiment using larger numbers of pigs in a more closely
controlled situation failed to produce any growth stimulation
due to a complete inability of the treatment to produce
specific antibody titres. This underlines the lack of basic
immunological knowledge which hampers such experiments.
Fundamental research on practical aspects of immunology
(conjugation methods, adjuvants, immunisation schedules, etc)
are required to allow this field to develop further.

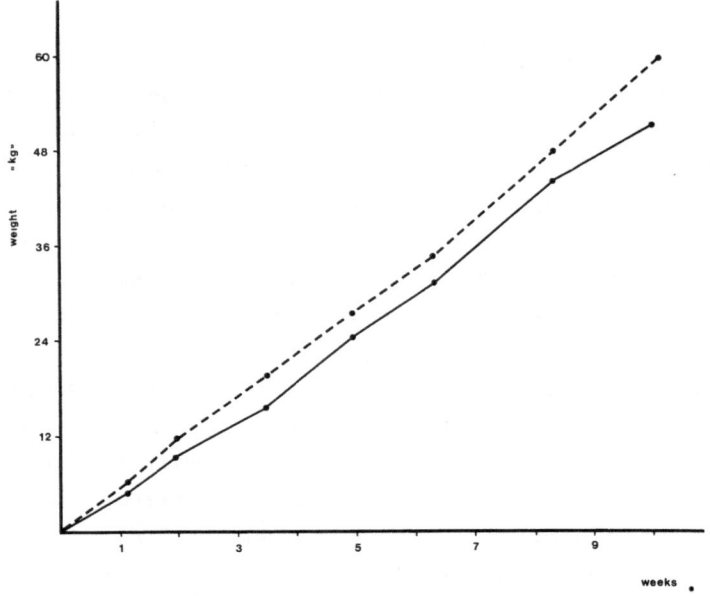

Fig. 3.   Weight gain (over initial weight) in Large White
pigs immunised against somatostatin (•-----•) and
control pigs immunised against androstenone (•——•).

There is therefore some evidence (albeit sketchy) that
this treatment works in different breeds and species and
under various conditions. With improved immunological
knowledge and techniques it should be possible to fully
evaluate the treatment as an effective commercial package.

Theoretically, this treatment has at least two major
advantages over conventional (anabolic steroid) growth
promotants. The effects can be greater than those reported
for implanted steroids - a 76% increase in body weight gain
having been recorded (Spencer and Williamson, 1981) while
anabolic steroids rarely produce more than a 20% increase
(Heitzman, 1978). The second advantage is that, with the use
of suitable adjuvants, there is not likely to be any problem
of residues to concern the consumer. The immunisation
treatment seems to be particularly effective during the early
growth period but the mechanism of action is still unclear.
Since somatostatin controls the release of such a wide
variety of hormones the effects of immunisation against
somatostatin may not be confined to growth alone. Other
aspects that may also be affected, such as wool growth or
milk production. Such alternative benefits are worthy of
further investigation and some such studies are currently in
progress.

CONCLUSION

It has been demonstrated that active immunisation of
animals against somatostatin can produce increased growth.
These experiments are, however, far from conclusive and
should only be viewed as indicative of a possible action.
They are practical demonstrations of a theoretically feasible
physiological mechanism and do not as yet provide anymore
than a basis for further investigations. The effect of this
treatment on diverse breeds and species needs to be fully
investigated, as does the mechanism of action. The
possibility also exists that this treatment may enhance milk
and wool production. Further investigations are currently
being undertaken but these, like commercial exploitation,
would benefit from increased fundamental immunological
knowledge of a practical nature.

ACKNOWLEDGEMENTS

The results reported here are from collaborative studies, initially with Dr. E.D. Williamson (Meat Research Institute, Langford, UK) and later with Dr. G.J. Garssen (Instituut voor Veeteeltkundig Onderzoek, Zeist, The Netherlands). Their contributions to this work are gratefully acknowledged.

REFERENCES

Arimura, A. and Fishback, J.B. 1981 Somatostatin: Regulation of secretion. Neuroendocrinology, 33, 246-256.

Bloom, S.R., Mortimer, C.H., Thorner, M.O., Besser, G.M., Hall, R., Gomez-Pan, A., Roy, V.R., Russel, R.C.G., Coy, D.H., Kastin, A.J., and Schally, A.V. 1974 Inhibition of gastrin and gastric acid secretion by growth hormone release inhibiting hormone. Lancet ii, 1106-1109.

Bloom, S.R., Ralphs, D.N., Besser, G.M., Hall, R., Coy, D.H., Kastin, A.J. and Schally, A.V. 1975 Effect of somatostatin on motilin levels and gastric emptying. Gut 16, 834.

Boden, G., Jacob, H.J., and Staus, A. 1976 Somatostatin interacts with basal and carbachol stimulated antral and duodenal motility. Gastroenterology 70, 961.

Daughaday, W.H., Hall, K., Raben, M.S., Salmon, W.D., Van den Brande, J.L. and Van Wyk, J.J. 1972 Somatomedin: Proposed designation for sulphation factor. Nature, 235, 107.

Ferland, L., Labrie, F., Jobin, M., Arimura, A. and Schally, A.V. 1976 Physiological role of somatostatin in the control of growth hormone and thyrotropin secretion. Biochem. Biophys. Res. Commun., 68, 149-156.

Gomez-Pan, A., Reed, J.D., Albinus, M., Shaw, B., Hall, R., Besser, G.M., Coy, D.H., Kastin, A.J. and Schally, A.V. 1975. Direct inhibition of gastric acid and pepsin secretion by growth hormone-release inhibiting hormone in cats. Lancet i, 888-890.

Grant, D.B., Hambley, J., Becker, D., and Pimstone, B.L. 1973 Reduced sulphation factor in undernourished children. Arch. Dis. Childhd. 48. 596-600.

Heitzman, R.J. 1978 The use of hormones to regulate the utilization of nutrients in farm animals: current farm practices. Proc. Nutr. Soc., 37, 289-293.

Koerker, D., Ruch, W., Chideckel, C., Palmer, J., Goodner, C., Ensinck, J. and Gale, C. 1974 Somatostatin: Hypothalamic inhibitor of the endocrine pancreas. Science, 184, 482-484.

Krulich, D., Illner, P., Fawcett, C.P., Quijada, M., McCann, S.M. 1972 Dual hypothalamic regulation of growth hormone secretion. In: "Proceedings of a Second International Symposium on Growth Hormone" (Ed. A. Pecile and E.E. Muller) (Excerpta Medica, Amsterdam) pp 306-316.

Machlin, J.L. 1976 Role of growth hormone in improving animal production. In: "Anabolic Agents in Animal Production" (Ed. F.C. Lu and J. Rendel) (G. Thieme, Stuttgart) pp 43-54.

Ruch, W., Koerker, D. Carino, M., Johnson, S., Webster, B., Ensinck, J., Goodner, C. and Gale, C. 1973 Somatostatin (somatotropin release inhibiting hormone) in conscious baboons. In: "Advances in Human Growth Hormone Research" (Ed. S. Raiti) (DHEW Publication NIH 74-612, U.S. Govt. Printing Office, Washington) pp 271-288.

Schusdziarra, V. 1980 Somatostatin - A regulatory modulator connecting nutrient entry and metabolism. Horm.

Schlegel, W., Harvey, R.F., Raptin, S., Oliver, J.M. and Pfeiffer, E.R. 1977 Inhibition of cholecystokinin-pancreozymin release by somatostatin Lancet ii, 166-168.

Spencer, G.S.G., and Williamson, E.D. 1981 Increased growth in lambs following auto-immunisation against somatostatin Anim. Prod. 32, 376.

Spencer, G.S.G., Garssen, G.J., Macdonald, A.A. and Colenbrander, B. 1982 Plasma somatomedin activity following insulin administration to fetal pigs. IRCS Med. Sci. 10, 1051-1052.

Spencer, G.S.G., Garssen, G.J. and Hart, I.C. 1983 A novel approach to growth promotion using auto-immunisation against somatostatin. I. Effects on growth and hormone levels in lambs. Livestock Prod. sci., 10 (in press).

Vale, W., Rivier, C., Brazeau, P. and Guillemin, R. 1974 Effects of somatostatin on the secretion of thyrotropin and prolactin. Endocrinology, 95, 968-977.

Varner, M.A., Davis, S.L. and Reeves, J.J. 1980 Temporal serum concentrations of growth hormone, thyrotropin, insulin and glucagon in sheep immunised against somatostatin. Endocrinology, 106, 1027-1032.

Wallis, W. 1980 The role of the pituitary gland in the regulation of somatomedin levels. In: "Proceedings of a Colloquium on Somatomedins". (Ed. G.S.G. Spencer) (Somatomedins Club, Bristol) pp 20-26.

DISCUSSION ON DR. SPENCER'S PAPER

Dr. Schanbacher: Could you comment on antibody titres?
Is there a relationship between the titres produced and the
growth response obtained? Also could you comment on the use of
passive immunization as opposed to active immunization.

Dr. Spencer: There is no correlation between antibody titres
and growth in the immunized lambs. If you have a measurable
antibody titre then you have excess antibody over that
required to inactive endogenous hormone production. As
regards the second question on passive as against active
immunization, we require growth stimulation throughout
the animals life time. Therefore, active immunization is the
more appropriate method to try.

Prof. Karg: How often were the animals injected?

Dr. Spencer: The animals were injected at 3 weeks
of age and fortnightly thereafter. This was our
experimental approach but this may not be necessary and in
fact we may have overdone it. The animals were eventually
failing to respond to stimulation as they were getting too
much.

Prof. Karg: Where was the injection site?

Dr. Spencer: The injection site was in the neck because
there is good drainage to the lymph nodes and also there are
chepaer cuts of meat there. We used Freunds complete
adjuvant which is very effective but not one that could
be used commercially, however.

Dr. Forbes: Were there any adverse reactions to immunization
in or on the skin of animals?

Dr. Spencer: In the first experiment we got large blisters
so in the second experiment we injected smaller doses at

each site and we then had no problems.

Prof. Hoffmann: Did you measure sex hormone levels in treated
and control animals?

Dr. Spencer: No.

Dr. Meyer: Did you immunize the control animals as well with
adjuvant as this causes stress?

Dr. Spencer: Yes. We probably have some depression in
growth as a result of the immunization. Maybe our
treated animals are growing no faster than animals not
immunized.

Dr. Galbraith: Are there differences in carcass deposition
of tissues because it is important to know whether there are
more dietary nutrients in lean meat and less in fat?

Dr. Spencer: The carcass is proportionally bigger.

Dr. Heitzman: I am not worried that you did not see changes
in growth hormone and insulin as exactly the same holds true
in sheep given combinations of anabolic agents which
maximise growth without changes in growth hormone. Do
you think the effects with immunization are additive to
increases obtained from anabolic agents?

Dr. Spencer: Yes, I do not see any reason why not as our
treatment is particularly effective in the young animals before
the epiphyseal plate closes. We do not have any evidence to
suggest anabolic agents act through the somatomedin axis.

Dr. Melrose: How many pigs were involved in the experiment?

Dr. Spencer: There were 6 in the treated group and 5 in the
controls.

Dr. Melrose: Have you tried immunization in improved
breeds of sheep?

Dr. Spencer: We are waiting for results of such trials.

Dr. Neimann Sorensen: Did you include both sexes and were
there any sex differences in your lamb work?

Dr. Spencer: There was no difference between male and
and female lambs so we pooled both groups.

Dr. Davis: We did a study similar to this but the
contradictory effects on growth may be explained by the
differences in methods and treatment. We saw an effect on
growth hormone secretion which was decreased after somatostatin immunization but we did not do a complete study on
growth. Our animals were older and larger. The immunization
approach in your study may have been done at a crucial
stage in the growth curve. Would you comment on both
studies.

Dr. Spencer: Yes. We killed our lambs at about the time
you started your experiment. Breeds were different
and our lambs would have reached puberty by then and
epiphyseal plate closure would have started. Since immunization
against somatostatin affects long bone growth, it is
important to start as early as possible. This even
suggests looking at pre-natal growth.

IMMUNOLOGICAL CASTRATION OF YOUNG BULLS FOR BEEF PRODUCTION

I.S. Robertson*, J.C. Wilson*, H.M. Fraser**,
G.M. Innes***, A.S. Jones***

*Royal (Dick) School of Veterinary Studies, University of Edinburgh
**M.R.C. Unit of Reproductive Biology, 37 Chalmers Street, Edinburgh
***The Rowett Research Institute, Aberdeen, Scotland

ABSTRACT

Successful immunological castration, through the neutralisation of
LHRH by active immunisation, has been achieved in young bulls. It results
in a period, about six months in length, of reduced testosterone secretion,
involution of the testes, azoospermia and docile behaviour. A slight
reduction in the rate of live-weight gain also occurs during this period,
but as LHRH antibody titres decline and the hypothalamus-pituitary-
testicular axis resumes functioning, the overall LWG is much superior to
steers (full castrates). Carcases of immunocastrates are also leaner than
steers although some degree of fat deposition occurs due to the temporary
reduction in testosterone secretion, which is also reflected in greater
long bone growth. It is suggested this technique be used to allow young
bulls, rather than steers, to be reared for beef, incorporating an immunocastration period on grassland in the management system.

INTRODUCTION

Neutralisation of luteinising hormone releasing hormone (LHRH) by
active or passive immunisation has been studied in a number of species (for
review see Fraser, Sharpe, Lincoln and Harmer, 1982). Immunoneutralisation
of LHRH by antibodies acting in the portal blood between the hypothalamus
and the pituitary gland prevents release of LH and FSH, resulting in a lack
of stimulation for testosterone secretion and spermatogenesis. In farm
animals, neutralisation of LHRH by active immunisation seemed to answer a
need for modification of male characteristics without loss of production
and other problems associated with conventional castration. In addition,
Short (1980) has postulated a growth and development advantage per se,
namely that temporarily castrated pre-pubertal young bulls could grow
larger skeletal frames due to continued long bone growth, similar to that
in fully castrated animals. A production benefit, however, could arise
from possible greater muscle formation at a later stage of development and
overall, could be superior to that of the normal bull.

We now describe four experiments in the order of their execution and
show how our concepts on the uses of immunological castration in cattle
developed. The results of Experiment I are published (Robertson, Wilson
and Fraser, 1979 and Robertson, Wilson, Fraser and Rowlands, 1981), those

of Experiment II are described here by courtesy of Glaxo Animal Health Ltd.,
London, and those of Experiment III are published (Robertson, Fraser, Innes
and Jones, 1982). Experiment IV is in progress.

EXPERIMENTAL PROCEDURES

British Friesian calves were used in all the experiments. The vaccine
was prepared by reacting synthetic LHRH with a protein carrier using carbo-
diimide as conjugating agent (see Robertson et al., 1979), and emulsifying
the resultant conjugate with an adjuvant. Administration was by subcutaneous injection of two 2 ml aliquots in the brisket or chest wall areas,
with each animal receiving approximately 0.1 mg LHRH. Two types of
adjuvant were used (1) Freund's complete (primary injection) and incomplete
(subsequent injections) or (2) an oil-in-water adjuvant supplied by Glaxo
Animal Health Ltd., London. A summary of procedures is shown in Table 1.
The methods for estimating antibody titres, serum testosterone levels and
testes volume have been described by Robertson et al., (1979).

TABLE 1   Summary of procedures used in four experiments.

Experiment	I	II	III	IV
No. calves immunised	2	5	10	16
Age at primary injection (weeks)	12	32	28	20
Control groups	nil	nil	10 steers	8 steers: 8 bulls
Period at grass	nil	nil	nil	4 months
Protein carrier	tetanus toxoid or porcine thyroglobulin (TG)	human serum albumin (HSA)	HSA	HSA (8) and TG (8)
Adjuvant	Freund	Glaxo	Freund	Freund

RESULTS AND DISCUSSION

Immune responses, semen production and behaviour

Experiment I with two treated animals was a pilot trial to determine
whether neutralisation of LHRH was feasible in male cattle. An effect
manifested by low serum testosterone levels and reduced testis volume was
obtained after antibody titres rose above 1 in 1000. However, four or more

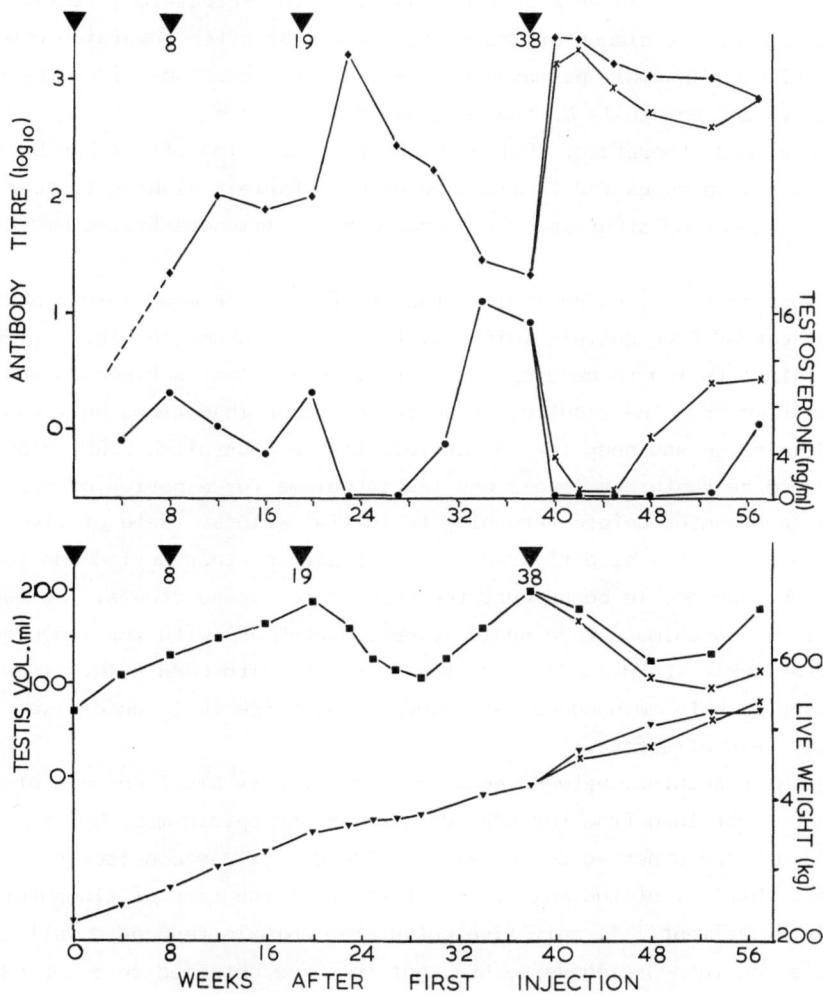

Fig. 1   Experiment II: Effects of immunisation (arrows) against
LHRH on antibody titre ◆ , on serum levels of testosterone ● ,
on testis volume ■ and on live weight ▼ when using immunogen containing human serum albumin throughout or porcine thyroglobulin
from week 38✕

injections at intervals were required. The effect persisted for six to
nine months, after which reversion to near normal parameters occurred.
During this period, the animals resembled steers in respect of their docile
behaviour and conformation. Semen collection and examination at intervals
showed that there was a marked depression of spermatogenesis, particularly

in one animal, correlated with the reductions in testosterone secretion and testis volume. At slaughter, however, six months after reversion commenced, mainly live and normal spermatozoa were readily recoverable from the vasa deferentia and the tails of the epididymes.

It seemed, therefore, that a temporary castration effect had been achieved and we suggested it might be used by farmers wishing to rear bulls, including a period at grass, when management of immunocastrates would be easier.

Experiment II yielded similar results to I. The mean values obtained in four out of five animals initially treated are shown in Fig. 1. The fifth animal failed to develop a satisfactory level of antibodies and has been omitted from the results. A series of three injections only was required to raise antibody levels sufficiently to neutralise LHRH, resulting in reduced testosterone levels and testis volume for a period of approximately four months before returning to initial values. Rate of live-weight gain was apparently slightly reduced for a part of this period and in appearance and docile behaviour, the animals resembled steers. Following reversion, two animals were again injected (week 38) with the immunogen used previously (LHRH + HSA), and the other two with LHRH + TG. The responses to both immunogens were good although the TG immunogen was slightly less effective.

After slaughter, between weeks 57 and 61, very small numbers of spermatozoa were obtained from the vasa deferentia and epididymes, but the accessory glands were observed to be well developed. It was considered unlikely, however, that any of the animals was fertile at the time of slaughter.

In Experiment III, only five out of ten animals responded well to immunisation (good responders, GR), but this was obtained in a series of two injections, an essential practical attribute. Their responses are shown in Fig. 2 in relation to the poor responders (PR) and, as before, in appearance and docile behaviour, they resembled steers during the period of low testosterone levels and testis volume which lasted approximately six months, an adequate length of time for grazing management. A third injection for the PR group in week 29 failed to elicit much response.

Semen samples taken from all the animals by artificial vagina or electroejaculation during week 29 contained occasional spermatozoa in only one animal per group. In week 41, one sample (PR) only contained a few spermatozoa, and in weeks 60 and 61 many dead, but no live, sperm were found in samples obtained within 30 minutes of slaughter from the tails of

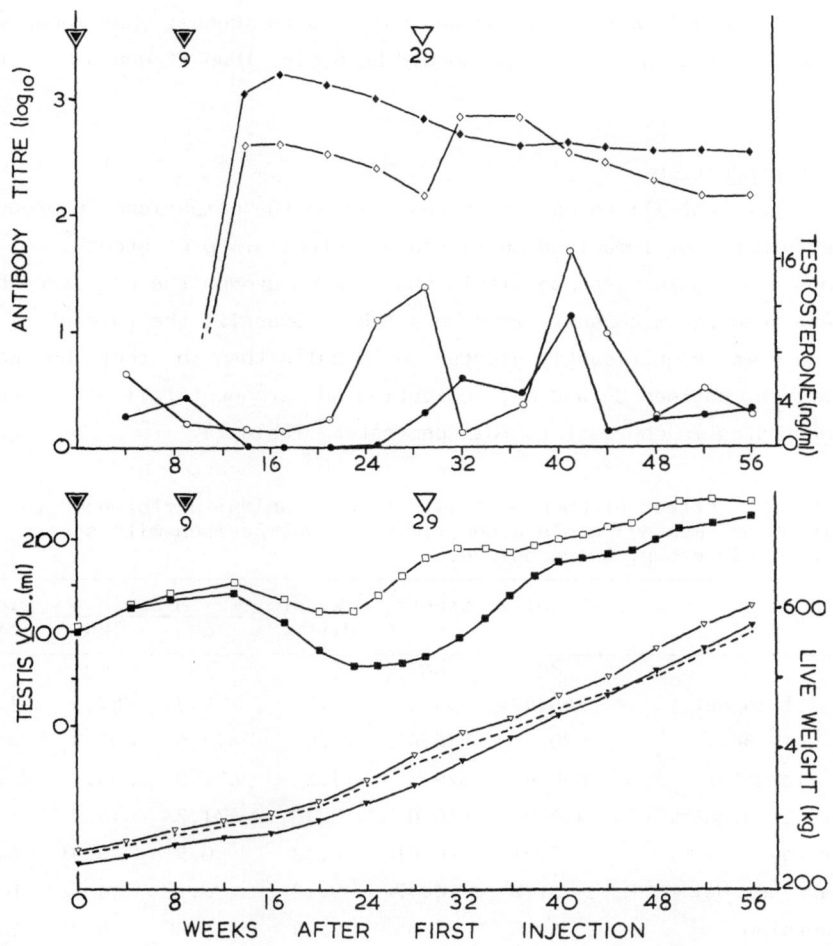

By courtesy of the Veterinary Record, London.

Fig. 2 Experiment III: Effects of immunisation (arrows) against LHRH on antibody titre in the good ◆ and poor responders ◇ , on serum levels of testosterone ● ○ , on testis volume ■ □ and on live weight ▼ ▽ . Liveweight in steers denoted with dotted lines. (All group means).

the epididymes in all of the animals.

Our current trial (Experiment IV) is still in progress. As before, half the number of animals have responded well. While the animals were on summer grazings, behavioural studies have shown that the GR group was similar to steers, being much less active than the PR and bull groups with regard to riding, heading, and licking and nosing. Although we have tested

142

this only in animals aged 8-11 months, the results suggest that management
of immunocastrated bulls at grass should be easier than it would be for
bulls.

Effect on production

In Experiment III we have also reported on the comparison in productive performance of immunised bulls and a control group of steers. As
there was a variable response within the treated group, the good responders
(GR) were compared with poor responders (PR). Overall, the rate of liveweight gain was significantly greater in IM bulls than in steers but was
not different between GR and PR, indicating no permanent fall-off in LWG in
immunocastrates in contrast to full castrates (Table 2).

TABLE 2   Effect of LHRH neutralisation on animal performance to
slaughter (means): whole group (IM bulls) comparison with steers
and within group, GR versus PR.

	IM Bulls	Steers	s.e. of diff.	IM Bulls GR	PR	s.e. of diff.
	kg	kg		kg	kg	
Wt. at 1st inject.	246.5	254.2	11.5	235.7	257.3	11.2[+]
Wt. at week 56	590.3	566.6	17.5	577.6	603.0	24.0
Wt. at slaughter	630.6	629.2	17.2	627.0	634.2	25.4
No. days to slghtr.	424.0	460.0	0.70[***]	431.2	416.0	1.70
LWG per day	0.91	0.81	0.32[**]	0.91	0.91	0.04
DM intake per day	6.22	6.37	0.12	6.12	6.32	0.16
DM conversion	6.88	7.85	0.24[***]	6.80	6.96	0.23
	cm	cm		cm	cm	
Withers ht. at slghtr.	134.9	137.9	1.64[+]	135.7	134.1	1.37
Foregirth at slghtr.	200.4	207.8	2.32[**]	200.2	200.6	2.90

Significance + P<0.10   Adapted from Robertson et al., 1982
             * P<0.05
            ** P<0.01
           *** P<0.001

As expected, the dry matter (DM) conversion rate followed the same
pattern as for LWG. Although the heavier initial weight of the PR group
(shown retrospectively) suggested that immunisation at an earlier stage
might be more effective, this was not confirmed in Experiment IV. With regard to conformation, IM bulls were shorter in height at the withers and

thinner around the foregirth than steers at the time of slaughter; GR animals were taller than PR.

In Table 3 data on some carcase traits are given. IM bulls are shown to be significantly different from steers in all of them and hence tend to resemble normal bulls. Within the IM bulls, however, the GR lean content of the 10th rib cut and eye muscle area was less than that of PR, and GR bone content was greater than PR. Cannon bone length was also greater in GR compared to PR.

TABLE 3 Comparison of some carcase traits between IM bulls and steers and between GR and PR groups (means).

| | IM Bulls | Steers | s.e. of diff. | IM Bulls | | s.e. of diff. |
				GR	PR	
Dressing out %	55.7	54.6	0.56[+]	55.8	55.7	0.69
Specific gravity of side	1.08	1.07	0.003[***]	1.08	1.08	0.003
Cannon bone length (cm)	22.5	23.5	0.30[**]	22.8	22.3	0.28[+]
10th rib cut:						
Lean %	55.6	45.1	1.8[***]	53.7	57.6	1.4[*]
Fat %	26.5	39.0	1.9[***]	27.8	25.2	1.6
Bone %	15.4	14.3	0.54[+]	16.3	14.5	0.69[*]
Eye muscle area (sq cm)	76.0	50.6	4.1[***]	71.5	80.4	5.52

Significance   +   P<0.10   Adapted from Robertson et al., 1982
                    *   P<0.05
                  **   P<0.01
                ***   P<0.001

These results on carcase composition show that production of lean meat in immunocastrates is much superior to that of steers. There appeared to be a slight fattening effect caused by the immunological castration which could be used to improve the usual very lean carcases of entire bulls and, hence, make them more acceptable in some markets. There was no evidence of a greater growth of muscle after reversion from immunological castration, although the greater withers height and cannon bone length of the immunocastrates may have been due to delay in epiphyseal fusion (Short, 1980). This latter effect presumably was a result of reduced testosterone secretion which is the generally accepted reason for extended growth in height in man and animals (Short, 1980). We believe that the effect obtained in relation to both bone growth and fat deposition following a temporary depression in endogenous testosterone secretion in cattle, has not been recorded pre-

viously. It appears that if cattle are deprived of the anabolic effects
of androgens, even temporarily, there is some diversion of energy intake
towards synthesis of fat. Whether this diversion in immunocastrates ceases
on reversion to full masculinity, is an interesting point, yet unclarified.

Residues in meat and carcase

We believe it unlikely that any significant level of antigen would
remain in the meat of immunologically castrated bulls, either shortly after
or, more usually, six to nine months after the last injection. If it were,
then it is unlikely to be harmful since LHRH is inactive when taken orally.
It is a naturally occurring peptide and is in widespread use in human medicine.

The circulating antibodies should also be harmless as they decline in
significance as reversion takes place but, in any event, these antibodies,
similar to other proteins, would be partially denatured in cooking and totally denatured by the digestive enzymes of the human consumer.

The subcutaneous nodules formed at injection sites have not prevented
carcases from being approved by meat inspectors who have, where necessary,
easily trimmed them off.

CONCLUSIONS

Our results indicate that immunological castration in young bulls has
a number of advantages:
1)   as a substitute for conventional castration, it is less traumatic and
     more humane;
2)   during the period of immunocastration, bulls are more tractable and
     highly unlikely to be fertile;
3)   it facilitates the rearing of bulls in management systems incorporating
     a period of grazing;
4)   immunocastrated bulls are more productive than steers, having superior
     growth rate and yield of lean meat.

Before the technique can become accepted practice, a controlled pre-
dictable response to vaccination must be guaranteed. A practical necessity
is a good response after a primary and secondary injection only, resulting
in a period of about four to six months temporary 'castration'. We have
been able to demonstrate this type of response but were unable to achieve
it in all treated animals due to variations in individual immune response
systems, and suboptimal choice of dose, protein carrier or adjuvant. Since

Freund's complete adjuvant has the added disadvantage of interfering with the tuberculin test, an alternative adjuvant or an adjuvant-free technique needs to be found. Further work to improve upon our results is required but it will be costly, especially in countries where it is constrained by drug testing regulations, as in the United Kingdom.

REFERENCES

Fraser, H.M., Sharpe, R.M., Lincoln, G.A. and Harmer, A.J. 1982. LHRH antibodies: Their use in the study of hypothalamic LHRH and testicular LHRH-like material, and possible contraceptive applications. In "Progress Towards a Male Contraceptive" (Ed. S.L. Jeffcoate and M. Sandler). (John Wiley & Sons Ltd., London). pp. 41-78.
Robertson, I.S., Fraser, H.M., Innes, G.M. and Jones, A. 1982. The effect of immunological castration on sexual and production characteristics in male cattle. Veterinary Record 111, 529-531.
Robertson, I.S., Wilson, J.C. and Fraser, H.M. 1979. Immunological castration in male cattle. Veterinary Record 105, 566-567.
Robertson, I.S., Wilson, J.C., Fraser, H.M. and Rowlands, A.C. 1981. Further studies on immunological castration in male cattle. Veterinary Record 108, 381-382.
Short, R.V. 1980. The hormonal control of growth at puberty. In "Growth in Animals" (Ed. T.L.J. Lawrence). (Butterworths, London). pp.25-45.

DISCUSSION ON DR. ROBERTSON'S PAPER

Dr. Neimann Sorensen: Were there differences in antibody titres between good and poor responders?

Dr. Robertson: Yes, the titres in the poor responders did not rise high enough to get a sufficient response. The poor responders were very similar to bulls in aggressive and copulatory behaviour.

GENERAL   DISCUSSION

Dr. Roche: Is there any evidence for the presence of bound residue of zeranol?

Prof. Rico: We did not examine this but it seems from other data there are no bound residues because residue levels are under 2 ppb and this explains the total radioactivity.   I have not got data on metabolism of zeranol in different tissues.

Prof. Hoffmann: This question can be extended.   Is there evidence that covalently bound residues are present after oestradiol-17β, testosterone and progesterone?

Prof. Rico: The problem of bound residues with trenbolone relates to its use by implantation.   If you do use implantation it is not very difficult to find bound residues because the excretion of bound residues is slow.   There is binding to endogenous compounds and the turnover of the endogenous compounds is very slow.   Following implantation, you are deal-ing with secretion of free or conjugated compounds and a small part is bound to endogenous compounds.   With time you then have an accummulation of bound residues and at the end of the implantation period of 60 - 70 days you have one part free and three parts bound.   The reason for this is the accummulation of bound radioactivity starting from a very low level to increased accummulation due to very low turnover of bound residues. There are similar problems with some other veterinary drugs, and I think this is the case for oestradiol 17β, testosterone and progesterone.

Prof. Karg: Is covalently bound trenbolone irreversibly bound?

Prof. Rico: Yes, covalently bound trenbolone is irreversibly bound.

148

Prof. Karg: Is there a long acting effect due to some of these properties by which trenbolone is covalently bound?

Prof. Rico: Once trenbolone is covalently bound its biologic activity is lost. The only activity is when the compound is free or actually binding in the target species. While binding there could be a toxic effect on target species, but once binding is finished the bound product is not toxic to the consumer because the binding is irreversible.

Prof. Karg: What is the evidence that it is inactivated due to binding and there is not reversibility?

Prof. Rico: The evidence is coming from reports on many other drugs. Covalent binding is not reversible thermodynamically, and it is very difficult to find the reacted metabolites responsible for the binding.

Dr. Meyer: Is it still not possible that trenbolone residue is only bound to the androgen receptor because ether is a very bad extractor of androgen from the receptor?

Prof. Rico: I do not think so because the binding between androgen and its receptor is always reversible but covalent binding is not reversible. If you have covalent binding to receptors it is different and irreversible.

Dr. Buttery: Are you sure that covalent binding is not just the incorporation of a tritium atom which has been released on catabolism of trenbolone into a purine or pyrimadine?

Prof. Rico: It is possible that part of it is due to this but this is not bound residue but another endogenous compound extractable with a label. With trenbolone, a large part of this radioactivity is not due to radioactivity in endogenous compounds but it is related to metabolites of trenbolone. For example, the hydrolysis of urea gives CO_2 and H_2O, but this reaction is thermodynamically irreversible. Any organism biosynthesising urea directly from CO_2 and H_2O uses a cycle to circumvent the energy barrier.

Prof. Hoffmann: We can do exhaustive ether extractions so I do not think there was any receptor bound label left. In addition, the amount of bound compounds far exceeds the amount required to bind to the number of receptors known to exist in muscle. We can therefore exclude the receptor. It is not possible to identify these metabolites. Part of it may be a trienvi structure.

Prof. Karg: Questions to Dr. Roche

Prof. Hoffmann: I cannot see why anybody would use sex steroid hormones in bulls due to low weight gain obtained. National habits vary from country to country.

Dr. Roche: We must look at this question in a number of ways. Firstly in the case of oestradiol + progesterone implantation in bulls we got an increase in carcass weight of 10 - 14kg. which is a significant boost to beef production. Secondly the reduced aggressive behaviour allows these bulls to be managed much easier, although this is difficult to quantify economically, it is a major consideration in management of bulls, particularly at pasture. The third aspect related to the increase in carcass conformation which should make the carcass more valuable. The increase in fat score may also be desirable as there was insufficient fat cover on the untreated bulls for storage and chilling. The fourth aspect related to effects on meat quality due to an expected reduction in incidence of dark cutting meat because of reduced stress in implanted bulls prior to slaughter. So many different considerations need to be taken into account and it is possible we may hormonally castrate bulls in future to reduce behaviour and also increase growth.

Mr. Walters: The initial fast release of oestradiol from Compudose did lead to a fair share of reports of increased sexual activity in the field. We now wash the implant to remove surface crystals of oestradiol and the initial release is now greatly reduced and has taken away a lot of sexual problems but not all of it. The duration of payout of Compudose has been raised by Dr. Roche and we have trials where we have got a response from the 8th to 12th month, but not every trial is broken down by period.

Dr. Roche: It has been known since 1966 that silastic rubber implants do have surface crystals on the outside. The question is will surface washing at manufacture suffice or do implants have to be washed prior to insertion. As regards the duration of payout, I am reporting the results as we obtained them of 70 - 80 animals per treatment. We have further preliminary data that oestradiol - progesterone pellet implants give a significantly better daily gain from approximately days 100 to 200 compared to Compudose given on day 1 of trial.

Dr. Heitzman: You said you lost about 15% of implants.

Dr. Roche: I said we could not palapte them in the ear which may be different than actually losing them.

Dr. Heitzman: Would that affect your results in the third period?

Dr. Roche: Yes it may. Preliminary evidence suggests differences in growth between steers with or without palpable implants. It is difficult to assess loss rate because the implant can move in the ear and this would need to be done on the basis of blood levels of oestradiol.

ENDOCRINE REGULATION OF GROWTH IN RUMINANTS[1]

S.L. Davis,[2] K.L. Hossner and D.L. Ohlson
Department of Animal and Veterinary Sciences
University of Idaho
Moscow, Idaho 83843

ABSTRACT

Hormone regulation of growth involves multiple hormones acting in concert. It is currently believed that the principal hormones are growth hormone and the somatomedins. These hormones, however, would be of limited effect without the presence of thyroid hormones and, at puberty, gonadal steroid hormones. Normal growth is additionally dependent upon insulin and glucocorticoids. Interactions among these hormones and cellular mechanism of hormone action are still poorly understood. It appears that optimal secretion of growth hormone and the somatomedins is dependent on breed, age, sex and nutrient intake. Generally those variables which are related to enhanced long term growth hormone secretion are also associated with greater rates of growth. Therefore, it should be theoretically possible to artificially enhance growth rate of domestic animals by enhancing growth hormone secretion, injecting synthetically derived growth hormone or by alteration of skeletal muscle cell responsiveness to a given amount of hormone. Although very little is known about the latter area, it would ultimately seem to be the most attractive approach because of proposed inherent genetic limitations in cell responsiveness to hormonal stimulation. In addition, any of the approaches suggested above may be further limited if applied singly rather than in the presence of additional amounts of secondary hormones associated with growth.

INTRODUCTION

Hormonal regulation of growth (in this review, growth is defined as skeletal enlargement and protein accumulation in skeletal muscle) is extremely complex involving many different hormones, nutrient availability and multiple interactions between the endocrine system and genetic and environmental factors. Although most of these hormones have been identified and some of the interactions defined, specifics of hormonal mechanisms of action are poorly understood.

There are other excellent recent reviews which deal with current concepts of endocrine control of growth (Daughaday, 1981; Florini, 1983; Bauman et al., 1982b). Therefore, we have limited this review to an

---

[1] Idaho Agric. Exp. Sta. Pub. No. 82415.

[2] Current address: Department of Animal Science, Oregon State University, Corvallis, Oregon 97331.

overview of hormonal regulation of growth in general terms followed by a review of growth hormone secretion in ruminants and a few thoughts on the future of endocrine manipulation for the purpose of enhancing growth in ruminants.

GROWTH HORMONE AND SOMATOMEDINS

The pituitary gland was implicated in the control of growth at about the turn of the century by the studies of Caselli, who first described the effects of hypophysectomy on growth of dogs (reviewed by Cheek and Hill, 1974). Isolation of biologically active growth hormone (GH) from beef pituitaries by Li and Evans in 1944 resolved the question of whether a specific separate growth factor was produced by the pituitary gland. Growth hormone is now known to be a distinct pituitary hormone responsible for a wide range of metabolic and growth-promoting effects _in vivo_.

Metabolism of the three major metabolic fuels (carbohydrate, lipid and protein) is affected by GH. The effects of GH on these processes have been extensively studied and reviewed (de Bodo and Altzuler, 1957; Knobil and Hotchkiss, 1964; Merimee and Rabin, 1973) and will not be examined in detail here. In brief, GH acts as a counter-regulatory hormone to insulin in carbohydrate metabolism, inducing hyperglycemia and upon chronic exposure, permanent diabetes mellitus (Young, 1953). Chronic GH treatment stimulates lipolysis and inhibits lipogenesis (Fain, 1962,1968; Kuhlemeier and Trenkle, 1966). These processes serve to redistribute metabolic fuels to tissues requiring them (Bauman et al., 1982b).

Growth hormone also has important effects on protein and amino acid metabolism. Administration of GH stimulates amino acid uptake (Noall et al., 1957; Riggs and Walker, 1960) and their incorporation into skeletal muscle proteins (Kostyo and Knobil, 1959; Manchester et al., 1959) of hypophysectomized rats. The GH-treated animal displays reduced nitrogen excretion and an overall positive nitrogen balance (Russell, 1957; Wheatley et al., 1966; Davis et al., 1970). These effects are, of course, essential for new tissue synthesis and growth.

In addition to stimulating skeletal muscle growth and general metabolic changes, GH was long thought to directly stimulate skeletal growth in mammals. It is now known that the effects of GH on skeletal muscle and

cartilage are due to a group of GH-dependent polypeptides called somatomedins (SM). It is not known whether SM also mediates the effects of GH
on carbohydrate and lipid metabolism. The SM hypothesis of GH action was
first postulated by Salmon and Daughaday in 1957 who demonstrated that
rat costal cartilage exposed to GH in vitro did not display the expected
increase in radioactive sulfate incorporation into cartilage. Hypophysectomized rat serum also failed to stimulate cartilage growth in
vitro, but serum from hypophysectomized, GH-treated rats caused an
increased incorporation of radiosulfate into cartilage. These observations led to the hypothesis that a serum "sulfation factor," dependent on
GH, was directly responsible for the increase in cartilage growth
observed after GH administration in vivo.

Sulfation factor has since been shown to have a wide array of
effects on other tissues including skeletal muscle, adipocytes and
fibroblasts and has been renamed somatomedin (Daughaday et al., 1972).
The extensive literature on somatomedins has been reviewed at length (Van
Wyk et al., 1974; Pecille and Muller, 1976; Van Wyk and Underwood, 1978;
Giordano et al., 1979; Phillips and Vassilopoulou-Sellin, 1980a,b). The
interested reader is referred to these excellent reviews for a more
in-depth examination of research concerned with the somatomedins.

Somatomedins are hormones which possess three biological properties
(Van Wyk and Underwood, 1978): insulin-like activity in nonskeletal
muscles (e.g., glucose oxidation in adipose tissue), their production is
GH-dependent and they stimulate incorporation of sulfate into cartilage
proteoglycans. Thus far, a number of somatomedins have been isolated,
primarily from the laboratory rat and humans. Peptides which have
somatomedin activity have been discovered independently in several
laboratories studying seemingly unrelated phenomena. Peptides first
termed somatomedins were isolated from human serum by Hall (1972) and
Uthne (1973), who prepared a neutral peptide fraction which stimulated
the incorporation of radiosulfate into chick pelvic cartilage leaflets.
This peptide was termed somatomedin A (SM-A). A second, basic somatomedin (SM-C) has also been isolated from human serum by Van Wyk et al.
(1974).

Another group was studying serum peptides with insulin-like activity
which was not suppressed with antiserum to insulin. They termed this

activity nonsuppressible insulin-like activity (NSILA; Froesch et al., 1963). This serum fraction was shown to contain two peptides with NSILA activity, NSILA-I and NSILA-II. Rinderknecht and Humbel (1976,1978a,b) have purified these peptides and renamed them insulin-like growth factors I and II (IGF-I and IGF-II) to emphasize their close structural and biological homologies with insulin.

During the same period, another group was investigating serum factors which were required for cell growth _in vitro_. These studies led to the purification of peptides from calf serum (Pierson and Temin, 1972) which were termed multiplication-stimulating activity (MSA). Later, MSA was isolated from serum-free culture medium of a rat hepatocyte cell line (Dulak and Temin, 1973a,b).

All of these factors (IGF, MSA, NSILA) have been shown to be somatomedins, based on their biological activities. Experiments conducted in several laboratories have demonstrated that the confusing somatomedin nomenclature can now be simplified (Daughaday, 1982). IGF-I is identical to SM-C, which has a rat serum counterpart, basic somatomedin. IGF-II is structurally similar to SM-A and rat MSA. Thus, two distinct somatomedin populations exist, at least in the rat and human, a basic SM and a neutral SM. Preliminary evidence (Davis et al., 1981) suggests that sheep serum also contains basic and neutral fractions with SM immunoreactivity.

Somatomedins have been shown by a number of investigators to be secreted by the liver. Current evidence suggests that SM secretion is controlled by GH as well as other hormones. McConaghey and Sledge (1970) perfused normal rat livers, _in situ_, and demonstrated an increase in sulfation factor activity when the livers were exposed to GH. Francis and Hill (1975), using liver perfusion, confirmed the GH effect and established prolactin as another hormone which enhanced release of the sulfation factor. Schalch et al. (1979) measured IGF accumulation in liver perfusates by a competitive protein binding assay and found that GH and prolactin stimulated SM (IGF) release _in vitro_. Using puromycin as an inhibitor of protein synthesis, they demonstrated that SM release was partially blocked, but some SM accumulated in the perfusate, suggesting that the SM released was synthesized _de novo_ and mobilized from intracellular storage sites. Insulin has also been implicated as a stimulus for

SM production in perfused liver from hypophysectomized rats (Daughaday et al., 1976) and in rat liver organ cultures (Binoux et al., 1982). Schalch and co-workers (1979) demonstrated a multiple hormonal require- ment for maintenance of SM levels in the hypophysectomized rat. Adminis- tration of GH alone or in combination with cortisol or triiodothyronine elevated circulating SM levels to about 40-50% of normal. SM levels were normalized only when all three hormones were given together. Recent evidence suggests that human IGF-II secretion is less dependent on GH than IGF-I (Zapf et al., 1981; Hintz and Liu, 1981). Hence, a complex pattern of control is emerging for the SM peptides suggesting that their hepatic synthesis and release is under multiple hormonal control and that the liver plays a central role in the endocrine control of growth.

Somatomedins are now believed to mediate all or most of the effects of GH on growth processes. Receptors for somatomedins have been demon- strated in cartilage, skeletal muscle, liver, fibroblasts and placenta (reviewed by Daughaday, 1982). Because the amounts of pure somatomedin available for research are very limited, the role of these polypeptides in growth has been inferred primarily from clinical observations, studies of isolated tissues and observations of somatomedin levels after various experimental manipulations. To date, few studies have been published which examine the effects of somatomedin administration in the intact animal.

Somatomedins have been demonstrated to stimulate skeletal muscle growth and differentiation in vitro. Ewton and Florini (1980) showed that proliferation and amino acid uptake by myoblasts was stimulated by MSA and SM-A. These same investigators (1981) showed that SM-C, IGF-I and MSA were capable of stimulating an increase in myoblast fusion to form multinucleated myotubes. Ewton and Florini (1981) also observed that GH had no effect on the processes of myoblast proliferation, fusion or amino acid transport, but was effective in stimulating amino acid uptake in myotubes. This differential effect of SM and GH on muscle cells, has led them to suggest that early muscle growth is mediated by SM whereas GH sensitivity of skeletal muscle may not appear until after differentiation occurs.

Developmental changes in serum SM levels have been observed in humans by Bala et al. (1981). They reported that SM was very low in the newborn infant but progressively increased from birth to 9-11 years of age, at which time, there was a rapid elevation of serum SM which reached peak values in 13-15 year old adolescents. Females consistently showed slightly higher concentrations of immunoreactive SM than males. Immuno-reactive SM-C concentrations in GH-deficient children treated with GH was positively correlated with body size (indicated by height and weight) as well as chronological and bone age (Blethen et al., 1982).

The above studies provide suggestive evidence that the SM play a positive role in growth promotion. Direct evidence implicating SMs and growth has been provided by van Buul-Offers and Van den Brande (1979 and 1982) and Holder et al. (1981). Both groups using partially purified SM and hypopituitary (Snell dwarf) mice, observed an enhanced growth rate in response to 2-4 weeks of daily SM injections. The SM preparation used by van Buul-Offers and Van den Brande induced an increase in body length and weight, while Holder's group noted a significant increase in body weight and radiosulfate uptake into cartilage in vivo. In contrast to van Buul-Offers and Van den Brande's observations, Holder et al. (1981) did not observe an increase in heart, liver and kidney weights or tail growth velocity. These discrepancies are probably due to the use of different SM preparations. Both groups used relatively impure SM which contained a variety of polypeptides in addition to the full spectrum of SM peptides. As the use of these impure preparations provides suggestive, but not conclusive evidence that SMs stimulate growth in the intact animal, the recent report by Schoenle et al. (1982) was a welcome addition to the SM literature. Using highly purified SM (IGF-I), Schoenle and coworkers (1982) demonstrated an enhanced body weight gain, lengthening of the tibial epiphyses and cartilage DNA synthesis in hypophysectomized rats treated for 6 days by constant SM infusion. The effects of this SM preparation paralleled those seen with GH infusion. This experiment confirmed the earlier results obtained with impure SM preparations and provides strong support for the SM hypothesis of growth control.

In conclusion, GH and SM stimulate a plethora of metabolic changes in the intact animal. The direct effects of GH may be limited to post-differentiation stages of muscle development and to production of the SM by the liver. The somatomedins, in turn, are believed to directly stimulate growth by their effects on cartilage, skeletal muscle and viscera.

THYROID HORMONE

The principal metabolically active compounds produced by the thyroid are thyroxine (T4) and 3,5,3'-triiodothyronine (T3), collectively referred to as thyroid hormone. Protein binding is central to the transport, tissue distribution and metabolic clearance of thyroid hormone, which specifically binds to several plasma and nuclear proteins. Plasma proteins transport the hormone through the circulation and hormone activity is expressed through nuclear receptors (Cody, 1978; Samuels et al., 1982).

Several lines of evidence suggest that T3 is more active than T4, even though T4 is the predominant form secreted by the thyroid. Plasma proteins that transport thyroid hormone bind more T4 than T3, allowing free T3 to exert biological activity; thyroxine is deiodinated to T3 in peripheral tissues; T3 has a wider tissue distribution than does T4; and T3 has a higher affinity for nuclear binding sites than T4 (Martin, 1974; Eberhardt et al., 1980).

Thyroid hormone plays a dual role in stimulating oxidative metabolism and anabolic functions of cells (Mosier, 1981). In fact, it seems that thyroid hormone affects the growth, development and metabolism of virtually all tissues in mammals and the specific effects depend upon the stage of development and the type of tissue involved. Thyroid hormone regulates tissue oxygen consumption, mineral balance and the synthesis and metabolism of proteins, carbohydrates and lipids (Eberhardt et al., 1980). A characteristic feature of thyroid hormone insufficiency is the slowing of several metabolic processes and a sharp decrease in basal metabolic rate. A complete discussion of the myriad effects of thyroid hormone are beyond the scope of this review. Instead, emphasis will be placed on the proposed role of thyroid hormone in general body growth.

In mammals, thyroid hormone is essential for normal growth and development. The role of thyroid function in the fetus is as yet unclear. However, evidence demonstrating dependency of some fetal tissue growth and development on thyroid hormone does exist. Furthermore, there is no unequivocal evidence to support the proposition that the fetal thyroid is not required for growth (Mosier, 1981).

Normal growth in the postnatal individual is dependent on a euthyroid state. For example, thyroid hormone deficiency during the neonatal period results in severe retardation of many organ systems, in the growth and development of body weight, length and the central nervous system (Eberhardt, 1980).

It is doubtful that a single mechanism of action is responsible for thyroid hormone's role in growth. Thyroid hormone may exert direct effects stimulatory to growth, such as increasing overall hepatic and muscle protein synthesis (Goldberg et al., 1979) or it may have permissive effects via other hormones such as GH. For example, it has been reported that thyroid hormone plays a role in regulating pituitary GH production. Moreover, thyroid hormone may also regulate the effects of GH on epiphysial growth (Eberhardt et al., 1980). It is likely that this dual role of thyroid hormone in the enhancement of GH synthesis and its potentiation of localized growth promoting actions of GH represents a significant portion of its contribution to normal growth of an individual.

Recent evidence has shown that the growth failure and reduced serum SM concentration associated with hypothyroidism are probably not due to decreased binding of GH to its cellular receptor (Chernousek et al., 1982). This study also presented evidence that hypothyroid rats still possess the capability for SM production. It is possible that the growth failure of hypothyroidism may be caused by decreased GH secretion and subsequently diminished SM production. However, tissue responsiveness to SM may also be decreased in hypothyroidism and thyroid hormone may play a role in restoring "normal" tissue response to growth promoting agents.

GONADAL STEROID HORMONES

It is widely accepted that gonadal steroid hormones are responsible for increased growth rates which occur in most mammals at puberty. Generally, androgens are associated with increased musculature as well as skeletal growth. Estrogens, on the other hand, initially enhance growth rate, but eventually cause closure of the epiphyses (a cessation in long bone growth) and do not result in as much skeletal muscle enlargement as observed with androgens (Brody, 1945; Berg and Butterfield, 1976; Brazel and Blizzard, 1974). The mechanisms by which gonadal steroid hormones affect this increased growth rate are not well defined. One proposed mechanism is that gonadal steroids stimulate GH secretion. Supporting this suggestion are observations that plasma GH concentrations are higher in intact than in castrate males (Anfinson et al., 1975; Davis et al., 1977a; Galbraith et al., 1978; Kertz et al., 1982). Conversely, treatment of castrate males with androgen increased plasma GH concentrations (Davis et al., 1977a). Treatment with estrogens also increases plasma GH concentration in steers (Trenkle, 1970), wethers (Davis et al., 1977a; Muir et al., 1983), ovariectomized ewes (Davis and Borger, 1974) and in human males (Wiedemann et al., 1976). However, administration of physiologic amounts of estradiol to ovariectomized cows failed to influence plasma GH concentrations (Beck et al., 1976).

It has also been reported that gonadal steroids influence SM activity, androgens being stimulatory (Salmon et al., 1963; Weidemann, 1981) and estrogens exerting an inhibitory effect (Weidemann and Schwartz, 1972; Saenger et al., 1976; Weidemann et al., 1976) in humans. However, others have reported that moderately supra-physiological levels of estradiol and testosterone added in vitro had little effect on sulfate incorporation into cartilage, either in the presence or absence of serum (Phillips et al., 1975). To our knowledge, there are no reports on the relationship between gonadal steroids and either bioassayable or immunoassayable SM activity in blood of ruminants.

It has also been proposed that gonadal androgens and estrogens exert direct effects on skeletal muscle growth. Reports on the presence of specific androgen receptors in skeletal muscle cytosol support this suggestion (Snochowski et al., 1980; Michel and Baulieu, 1980). However, androgens have not been observed to be particularly effective in stimulating protein synthesis or cell division of skeletal muscle cultured in vitro (Gospodarowicz et al., 1976; Allen et al., 1983; Florini, 1983) although these workers did not determine if the cultured cells still contained androgen receptors. The postulated presence of estrogen receptors in skeletal muscle has not yet been documented.

OTHER HORMONES

Hormones in this category may be classified as serving a supportive role in growth. These include, in the opinion of the authors, glucocorticoids, insulin, parathyroid hormone, calcitonin and perhaps even prolactin (PRL). In their presence, growth may proceed at normal rates but they do not, by themselves, directly stimulate growth. In previous studies, it was observed that repeated injections of tyrotropin releasing hormone (TRH) stimulated secretion of TSH and PRL, and also increased growth rate in lambs and calves (Davis et al., 1976; Davis et al., 1977b). This observation has been confirmed in other studies with calves (McGuffey et al., 1977; P.W. Swift, personal communication). We have attributed the increased growth rate to the combined increase in blood levels of PRL and TSH since we did not see an effect of TRH on GH secretion. This interpretation is consistent with previously reported GH-like activity of PRL (Beck et al., 1964; Bauman et al., 1982b) and thyroid hormone effects on growth (van Buul-Offers and Van den Brande, 1979 and 1982).

Some may take issue with our classification of insulin as a secondary growth regulatory hormone. Indeed, insulin has been shown in vitro to stimulate amino acid uptake, protein synthesis (Wool et al., 1968) and perhaps decreased proteolysis (Williams et al., 1980). Such observations may have led to the suggestion that insulin is directly or primarily involved in muscle growth. However, in a recent review, Florini (1983) has observed that insulin is normally added to tissue culture medium at concentrations of 10^{-6}M, while it is generally present in the circulation at concentrations of 10^{-9} to 10^{-10}M. In view of the cross reactivity of insulin and SM at the receptor level (Ewton and Florini, 1981), it has been proposed that the observed in vitro effects of insulin on protein synthesis in cultured muscle cells is due to insulin binding to SM receptors. Insulin also has an apparent indirect effect on growth by enhancing SM secretion (Schalch et al., 1979). Although this issue remains to be resolved, we continue to think of insulin as playing a secondary or supportive role in muscle growth regulation and perhaps a primary role in initiating a redirection of energy into fat stores. It has recently been reported that the plasma concentration of insulin/unit metabolic body weight (BW$^{.75}$) increased (Table 1) above body weight of

230 kg in cattle (Kertz et al., 1982). At the same time, the GH to BW$\cdot^{75}$ ratio declined. In view of the lipogenic activity of insulin and lipolytic effects of GH, it would seem reasonable to speculate that these trends favor a decreased rate of muscle protein synthesis and an increased rate of fat deposition.

In the authors' opinion, GH and SM are likely the primary growth regulating hormones. Therefore, the subsequent sections of this paper will deal principally with current information on these two hormones and their relationship to growth in ruminants.

TABLE 1. Relationship between body weight and the ratios of plasma insulin (μU/ml) and GH (ng/ml) per unit of metabolic body weight (BW in kg$\cdot^{75}$).[a]

Ratio	Body Weight (kg)			
	187	230	310	376
Insulin/BW$\cdot^{75}$	.427	.305	.417	.439
GH/BW$\cdot^{75}$	.202	.210	.129	.104

[a] Adapted from Kertz et al. (1982) with the authors' permission.

PATTERNS OF GH SECRETION IN RUMINANTS

Although radioimmunoassay technology for GH in ruminants has been available for the past 15 years, early attempts to define a relationship between circulating concentrations of GH and the physiology of growth were of limited success. This limitation was due, in large part, to an early assumption that GH concentrations in blood would be relatively static and that blood concentrations would be proportional to the physiologic measure of interest. Studies in the past 10 years have clearly demonstrated that the assumption of static GH secretion was incorrect. Instead, GH secretion and blood GH concentrations are quite dynamic over time in humans, rats, cattle and sheep (McAtee and Trenkle, 1971; Davis and Borger, 1974; Anfinson et al., 1975; Finklestein et al., 1972; Tannenbaum and Martin, 1976) with the spikes or peaks of GH secretion occurring apparently at random in ruminants (Anfinson et al., 1975; Davis and Borger, 1974; Davis et al., 1977a; Figure 1). This

162

pattern of secretion has made it extremely difficult to assess a relationship between GH concentration and physiologic function. Therefore,
one cannot merely obtain single blood samples for GH measurement and
expect that single estimate to be an accurate representation of actual
concentration. Rather, it is necessary to obtain multiple, sequential
blood samples over time (a minimum of 6 hours in our opinion). Furthermore, analysis of such data presents additional problems, particularly if
one is interested in examining more than just average GH concentrations.
Since GH is secreted episodically, we believe it is reasonable to assume
that the pattern of secretion has physiological relevance. In other
words, from a physiologic standpoint, there is particular significance to
the frequency of the secretory spikes (number/unit time), the amplitude
of the secretory spikes (average maximum value) and the baseline GH

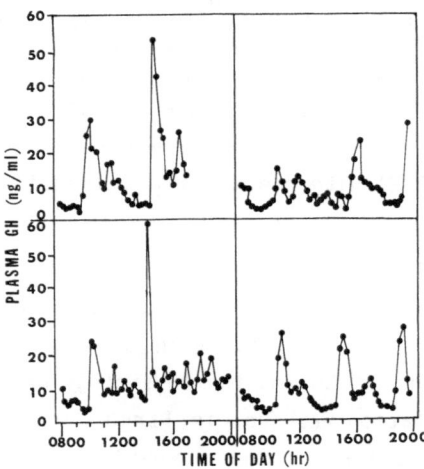

Figure 1. Plasma concentrations of GH in four post-
pubertal rams. Blood
samples were obtained at 15
minute intervals over 12
hours. (From Davis et al.,
1977a. Reproduced with
permission.)

concentration (average of all observations minus those which are part of
a spike). This suggestion is supported by studies reported recently
where injection of GH four times a day to hypophysectomized rats induced
a greater growth response than two injections per day (Jansson et al.,
1982a). Similar results were reported in another study where GH was
injected either 1, 2, 4 or 8 times/day (Jansson et al., 1982b). Further
support of the physiologic relevance of episodic hormone secretion is

found in reports that such patterns are essential for normal function of protein and steroid reproductive hormones (Desjardins, 1981; Brinkley, 1981; Knobil, 1981). For example, gonadotropin secretion is inhibited by continuous infusion of gonadotropin-releasing hormone and that inhibition is reversed when a pulsatile treatment pattern is instituted (as reviewed by Knobil, 1981).

However, in a study with calves, treatment with GH by continuous infusion or multiple injections both increased nitrogen retention, but no difference in response was noted between the modes of GH treatment (Moseley et al., 1982). These results don't rule out the possibility that a greater response may have been observed had a different pulse pattern of GH administration been used.

At any rate, the assumption that the pattern of hormone secretion has physiologic relevance has led us and others to develop methods designed to provide some degree of objectivity in estimating these different variables of hormone secretion (Santen and Bardin, 1974; Christian et al., 1978; Van Cauter et al., 1981; Merriam and Wachter, 1982) to allow comparisons of spike amplitude and frequency, and baseline hormone concentrations between treatments, groups or individuals.

Influence of gonadal steroids on GH secretion in ruminants

As mentioned in a previous section, estrogens and androgens stimulate GH secretion. For example, we observed that castration resulted in decreased GH baseline and spike amplitude in lambs and calves (Anfinson et al., 1975; Davis et al., 1977a) while treatment of wethers with diethylstilbestrol (DES) or testosterone (T) resulted in increased baseline GH concentration but had no significant effect on GH spike amplitude (Table 2; Davis et al., 1977a). In an earlier study with ovariectomized ewes (Davis and Borger, 1974), we noted that estradiol-17β also enhanced GH secretory activity. Therefore, differences in GH concentrations between castrate and intact ruminants are likely the result of gonadal steroid removal. We have also noted, however, that

164

TABLE 2   Variables of GH secretion in calves and lambs, both intact and castrated, with and without gonadal steroid hormone treatment

Variable	Calves[a]		Sheep[b]			
	Bulls	Steers	Ram	Wether	+T	+ DES
Overall mean, ng/ml	7.0	5.7	10.2	3.6	5.8	6.0
Baseline, ng/ml	5.6	5.1	8.8	2.7	4.9	5.3
Amplitude, ng/ml[c]	16.8	10.6	24.4	7.4	12.7	12.4
Frequency, no./hour	.4	.3	.2	.4	.2	.2

[a] From Anfinson et al., 1975.
[b] From Davis et al., 1977a.
[c] Not corrected for baseline.

baseline and spike amplitude are higher (Table 3) in ram lambs than in ewe lambs. A similar sex difference has also been reported for cattle (Reynert et al., 1976; Kertz et al., 1982). These observations may suggest a greater inherent capability to secrete GH in males, or that testicular steroids are more stimulatory to GH secretion in ruminants than are ovarian steroids. As Reynert et al. (1976) also observed this difference in pre-pubertal calves, gonadal steroids may not be involved.

TABLE 3   Variables of GH secretion in post-pubertal ram and ewe lambs[a]

Variable	Ram	Ewe
Body weight, kg	60.0	42.1
Age, mo	7.7	7.2
GH		
Overall mean, ng/ml	2.6	1.2
Baseline, ng/ml	2.4	1.1
Amplitude, ng/ml	6.6	2.7
Frequency, no./hour	.5	.3

[a] Davis, S.L. and Ohlson, D.L., unpublished data.

Relationship between breed and GH secretion

Different breeds and breed crosses in cattle and sheep exhibit markedly different growth rates. However, the relationship between genetic differences in growth and hormone secretory ability are poorly understood.

In a study of progeny from Angus X Hereford dams and either Charolais or Angus sires, measurements of GH status were positively related to carcass muscle and RNA in muscle, but breed of sire had no effect on any endocrine measurement (Trenkle and Topel, 1978). More recently, significant breed effects on plasma GH have been reported. For example, Keller et al. (1979) noted that serum GH concentrations were greater in Angus than in Hereford cattle at all ages studied. Similarly, Ohlson et al. (1981) observed a significantly higher overall plasma GH concentration in Simmental as compared to Hereford bull calves of the same age (Table 4). This higher overall average GH concentration was due to significantly higher baseline GH concentrations and numerically higher average GH spike amplitude. In crossbred steers of differing growth potentials, average plasma GH was higher (7.3 ng/ml) in large breed type than in small breed type (5.5 ng/ml) animals (Verde and Trenkle, 1982). In contrast, Kertz et al. (1982) reported that GH concentrations were poorly associated with breed. However, these workers had age confounded with breed and analyzed single blood samples per animal.

TABLE 4   Measure of GH secretion in Simmental and Hereford bull calves[a]

Variable	Hereford	Simmental	P
Overall, ng/ml	3.5 + .4	5.2 + .4	P<.01
Baseline, ng/ml	3.2 + .3	4.9 + .5	P<.05
Amplitude, ng/ml[b]	3.4 +1.0	5.0 +1.3	NS[c]
Frequency, no./hour	.2	.2	NS

[a] From Ohlson et al., 1981
[b] Adjusted amplitude (i.e., average peak height minus average baseline concentration.
[c] P>.05.

An experiment has been recently conducted in which GH secretion and growth rates in Targhee ram lambs selected for superior growth rate and feed efficiency were compared with non-selected Targhee ram lambs (Dodson et al., 1983). Overall mean GH concentrations were higher in selected

than in unselected lambs (Table 5). We have attributed this difference
to higher GH spike amplitude even though these were not statistically
significant because the baseline GH concentrations were essentially
identical between genetic lines.

TABLE 5   Growth rates and measures of GH secretion in Targhee
rams.[a]

Variable	Unselected	Selected	P
Gain/day age, kg	.181	.205	P<.05
Overall GH, ng/ml	4.6 + .5	6.1 + .4	P<.02
Baseline GH, ng/ml	3.2 + .5	3.7 + .3	NS[c]
Amplitude, ng/ml[b]	7.4 + .8	12.1 + 3.0	NS
Frequency, no./hour	.4	.5	NS

[a] From Dodson et al., 1983
[b] Average amplitude minus baseline.
[c] P>.05.

These results suggest that there is a greater ability to secrete GH
in more rapidly growing breeds or lines within breed. We further suggest
that this relationship may be causal in nature. These observations and
the report that within animal repeatability of GH secretory ability is
reasonably high (.79; Davis et al., 1979) suggest that such measures of
GH secretion may be of value in predicting genetic growth potential.

Nutritional influences on GH

It has been reported that plasma or serum GH concentrations increase
with fasting and decrease with feeding (Trenkle, 1978; Bassett, 1974a,b;
McAtee and Trenkle, 1971). These changes in GH concentrations, however,
could be due either to increased GH secretion or decreased GH clearance
from the bloodstream. In a study with calves and sheep, fasting was
associated with a reduction in the metabolic clearance rate (MCR) of GH
(Trenkle, 1976). Therefore, even though plasma GH concentrations may tend
to increase during fasting, this increase may be due partially to
decreased MCR from the circulation. However, it does appear that short
term changes in GH concentrations do occur in ruminants in association

with feed intake (McAtee and Trenkle, 1971; Driver and Forbes, 1980; Blom et al., 1976; Bassett, 1974a,b; Hove and Blom, 1973). Generally, meal consumption was associated with an immediate decline in plasma GH, followed within minutes with increased GH in cattle (McAtee and Trenkle, 1971) and sheep (Bassett, 1974a,b). However, the exact nature of the GH response to feeding in sheep appeared to be age related since the initial decrease persisted longer in mature animals and the eventual rise was less than in preweaned lambs (Bassett, 1974a).

Diurnal variation

There do not appear to be consistent diurnal fluctuations in plasma GH concentrations in ruminants. This observation is in contrast to a sleep associated increase in plasma GH in humans (Finkelstein et al., 1972) and a definite diurnal pattern of plasma GH in rats (Tannenbaum et al., 1976). In our laboratory, no diurnal pattern was noted in plasma GH in cattle (Anfinson et al., 1975). Similarly, no consistent diurnal pattern was observed in sheep or goats (Driver and Forbes, 1980; Tindal et al., 1978). In contrast, Hove and Blom (1973) have reported a diurnal pattern of plasma GH in dairy cows, but they attribute these changes to feed consumption.

Somatomedins (SM) in ruminants

Using bioassay methods, several investigators have examined SM activity in blood from ruminants. Using $^{35}SO_4$-uptake into costal cartilage, Van den Brande et al. (1974) reported that interspecies (including bovine) responses to somatomedin are far less restricted than interspecies responses to GH.

By definition, SMs are peptides which are GH dependent. In other words, SM production in most instances is proportional to GH secretion. The evidence for such a relationship in ruminants is limited. In fact, Falconer et al. (1977) have reported that bioassayable SM in ewes decreased in response to hypophysectomy but did not increase in response to GH treatment. Olson et al. (1981), on the other hand, have reported a 5-fold increase in radioimmunoassayable (RIA) SM activity in a single ewe with a GH secreting tumor while Underwood et al. (1982) noted a decrease in serum SM (by RIA) from 2.4 in normal serum to .5 U/ml in serum from hypophysectomized sheep. Therefore, SM activity in sheep appears to be positively related to if not dependent on GH. Somatomedin activity (chick pelvic leaflet assay) has been reported to increase between 6 and 10 mo and be positively correlated to rate of gain in bull calves (Lund-Larsen et al., 1977; Ringberg, 1979). Falconer et al. (1980) noted higher plasma SM activity (porcine costal cartilage assay) in Friesian than in Angus X Friesian bull calves. These apparent breed effects in calves are consistent with higher serum SM activity (chick epiphyseal cartilage assay) reported for faster growing Suffolk sired lambs than in slower growing Finnsheep sired lambs (Wangsness et al., 1981). Furthermore, these same investigators have reported that relative body weight gains in sheep are positively correlated to serum SM activity (Olsen et al., 1981).

The bulk of these data suggest that measures of SM activity in ruminants are related to growth rate and plasma GH. However, bioassay systems are non-specific and exhibit considerable variability (C.E. Powell and S.L. Davis, unpublished data). Therefore, in the authors' opinion, the precise nature of these relationships and the value of blood SM concentrations as a predictor of growth in ruminants must await development of more specific, accurate and precise radioimmunoassays for ovine and bovine SM.

PROPOSED APPROACHES TO GROWTH STIMULATION

Stimulation of GH secretion

Recent advances have made it feasible to administer substances which specifically stimulate the secretion of GH. Bowers et al. (1980) have reported the effects of a synthetic enkephalin analog on the secretion of GH in rats. Using a similar analog, we observed that plasma GH was increased in a dose-dependent manner in lambs (S.L. Davis and D.L. Ohlson, unpublished data). More recently, Guillemin et al. (1982) have described the structure of a GH releasing factor isolated from a pancreatic tumor of an acromegalic human male. This peptide also increased GH secretion in rats. Data are not yet available on the influence of this peptide on pituitary hormone secretion in domestic animals. Furthermore, there are essentially no data in the literature on the ability of peptides which stimulate GH secretion to also increase growth rate in domestic animals. It would also be of interest to examine the growth effects of these GH releasing peptides in combination with other substances which reportedly increase growth rate such as thyrotropin-releasing hormone (Davis et al., 1976; McGuffey et al., 1977), estrogens and trenbolone acetate (Galbraith and Geraghty, 1982; Heitzman, 1981).

Administration of hormones synthesized by genetic engineering techniques

Genetic engineering companies have recently demonstrated the technical feasibility of producing pig and cattle GH using bacteria containing the appropriate gene. Recombinantly derived bovine GH (bGH) was as effective in stimulating lactation as bGH purified from pituitaries (Bauman et al., 1982a). The ability of exogenously administered GH to stimulate growth is less clear. While some have demonstrated that exogenous GH increases growth or nitrogen retention (Davis et al., 1970; Machlin, 1972; Moseley et al., 1982) others have observed no beneficial effect on growth (Muir et al., 1983). Some of these apparent contradictions could be due to the pattern of GH administration (continuous vs single daily injections vs multiple injections [see discussion in a previous section]).

In view of the apparently central role of somatomedins in growth regulation, and the observed stimulation of body growth with somatomedins (Schoenle et al., 1982; van Buul-Offers and Van den Brande, 1979 and 1982; Holder et al., 1981), it is tempting to speculate that somatomedin administration may also be capable of enhancing growth rate in domestic animals. Unfortunately, sufficient amounts of any somatomedin are not available for such studies in domestic animals. Furthermore, the chemical natural and biologic properties of somatomedins from pigs, cattle, sheep and chickens are not well defined. We have recently obtained small amounts of purified sheep somatomedin and it appears to be chemically (Davis et al., 1981) and immunologically (Underwood et al., 1982) similar to human somatomedin-C. Considerable work remains, however, before the degree of similarity or chemical homology with human SM-C is determined.

Insertion of GH gene into embryos

As genetic manipulation of mammalian embryo genes is perfected, it should be possible to insert the GH gene directly into embryos. It has already been demonstrated that the gene for rabbit hemoglobin or for human β globin can be successfully inserted into fertilized mouse eggs. The human GH gene has also apparently been inserted into mouse eggs with some degree of success (Mintz as cited by Marx, 1982). Whether or not such "engineered genes" will be translated into an increased ability to secrete GH and subsequently to stimulate improved growth still remains speculative. The results will undoubtedly be interesting regardless of the outcome.

Manipulation of cellular responsiveness to hormonal stimulation

In our opinion, the greatest potential for increasing growth and production efficiency in domestic animals is at the cellular level. Administration of additional amounts of GH or substances which stimulate GH secretion may be effective only if the animals' cells are capable of recognizing and responding to such treatment. Hormonal activation of the intracellular biochemical events may be rate limiting depending on the genetic make-up of the animal. Unfortunately, this is the area about which we currently know the least.

A word of caution is in order concerning any of these suggested means of growth modulation. An array of hormones, as reviewed in the first section of this manuscript, act in concert to regulate growth. It is unlikely, then, that any single approach to hormonal manipulation will achieve optimal growth response.

REFERENCES

Allen, R., Masak, K.C., McAllister, P.K. and Merkel, R.A. 1983. Effect of growth hormone, testosterone and serum level on actin synthesis in cultured satellite cells. J. Animal Sci. 56: In press.
Anfinson, M.S., Davis, S.L., Christian, E. and Everson, D.O. 1975. Episodic secretion of growth hormone in bulls and steers: An analysis of frequency and magnitude of secretory spikes occurring in a 24-hour period. Proc. Western Section American Society of Animal Science. 26: 175-177.
Bala, R.M., Lopatka, J., Leung, A., McCoy, E. and McArthur, R.G. 1981. Serum immunoreactive somatomedin levels in normal adults, pregnant women at term, children at various ages and children with constitutionally delayed growth. J. Clin. Endocrinol. Metab. 52: 508-512.

Bassett, J.M. 1974a. Diurnal patterns of plasma insulin, growth hormone, corticosteroid and metabolite concentrations in fed and fasted sheep. Aust. J. Biol. Sci. 27: 167-181.

Bassett, J.M. 1974b. Early changes in plasma insulin and growth hormone levels after feeding in lambs and adult sheep. Aust. J. Biol. Sci. 27: 157-166.

Bauman, D.E., DeGeeter, M.J., Peel, C.J., Lanza, G.M., Gorewit, R.C. and Hammond, R.W. 1982a. Effect of recombinantly derived bovine growth hormone (bGH) on lactational performance in high yielding dairy cows. J. Dairy Sci. 65: Supp. 1, 121.

Bauman, D.E., Eiseman, J.H. and Currie, W.B. 1982b. Hormonal effects on partitioning of nutrients for tissue growth: role of growth hormone and prolactin. Federation Proc. 41: 2438-2544.

Beck, J.C., Gonda, A., Hamid, A., Morgan, R.O., Rubenstein, D. and McGarry, E.E. 1964. Some metabolic changes induced by primate growth hormone and purified ovine prolactin. Metabolism. 13: 1108-1134.

Beck, T.W., Smith, V.G., Seguin, B.E. and Convey, E.M.. 1976. Bovine serum LH, GH and prolactin following chronic implantation of ovarian steroids and subsequent ovariectomy. J. Animal Sci. 42: 461-468.

Berg, R.T. and Butterfield, R.M. 1976. Factors affecting muscle growth patterns. In "New Concepts of Cattle Growth" (Sydney University Press). pp. 99-142.

Binoux, M., Lassarre, C. and Hardouin, N. 1982. Somatomedin production by rat liver in organ culture. III. Studies on the release of insulin-like growth factor and its carrier protein measured by radioligand assays. Effects of growth hormone, insulin and cortisol. Acta Endocrinol. 99: 422-430.

Blethen, S.L., Daughaday, W.H. and Weldon, V.V. 1982. Kinetics of somatomedin C/insulin-like growth factor I: Response to exogenous growth hormone (GH) in GH-deficient children. J. Clin. Endocrinol. Metab. 54: 986-990.

Blom, A.K., Halse, K. and Hove, K. 1976. Growth hormone, insulin and sugar in the blood plasma of bulls. Interrelated diurnal variations. Acta Endocrinol. 82: 758-766.

Bowers, C.Y., Momany, F., Reynolds, G.A., Chang, D., Hong, A. and Chang, A. 1980. Structure-activity relationships of a synthetic pentapeptide that specifically releases growth hormone in vitro. Endocrinol. 106: 663-667.

Brazel, J. and Blizzard, R.M. 1974. The influence of the endocrine glands upon growth and development. In "Textbook of Endocrinology" (Ed. R.H. Williams). (W.B. Saunders Company, Philadelphia). pp 1031-1037.

Brinkley, H.J. 1981. Endocrine signaling and female reproduction. Biol. Reprod. 24: 22-43.

Brody, S. 1945. "Bioenergetics and Growth." (Reinhold Publishing Corp., New York). pp. 162-168.

Cheek, D.B. and Hill, D.E. 1974. Effects of growth hormone on cell and somatic growth. In "Handbook of Physiology," Sect 7: Endocrinology, v. 4, The Pituitary Gland (Ed. R.O. Greep and E.B. Astwood). (American Physiological Soc., Washington, DC). pp. 159-185.

Chernausek, S.D., Underwood, L.E. and Van Wyk, J.J. 1982. Influence of hypothyroidism on growth hormone binding by rat liver. Endocrinol. 111: 1534-1538.

Christian, L.E., Everson, D.O. and Davis, S.L. 1978. A statistical method for detection of hormone secretory spikes. J. Animal Sci. 46: 699-706.

172

Cody, V. 1978. Thyroid hormones: crystal structure, molecular conforma-
    tion, binding and structure-function relationships. Rec. Prog.
    Horm. Res. 34: 437-475.
Daughaday, W.H. 1981. Growth hormone and the somatomedins. In "Endocrine
    Control of Growth" (Ed. W.H. Daughaday). (Elsevier, New York). pp.
    1-24.
Daughaday, W.H. 1982. Divergence of binding sites in vitro action and
    secretory regulation of the somatomedin peptides, IGF-I and IGF-II.
    Proc. Soc. Exp. Biol. Med. 170: 257-263.
Daughaday, W.H., Hall, K., Raben, M.S., Salmon Jr., W.D., Van den Brande,
    J.L. and Van Wyk, J.J. 1972. Somatomedin: Proposed designation for
    sulphation factor. Nature, 235: 107.
Daughaday, W.H., Phillips, L.S. and Mueller, M.C. 1976. The effects of
    insulin and growth hormone on the release of somatomedin by the
    isolated rat liver. Endocrinol. 98: 1214-1219.
Davis, S.L. and Borger, M.L. 1974. Dynamic changes in plasma prolactin,
    luteinizing hormone and growth hormone in ovariectomized ewes. J.
    Animal Sci. 38: 795-802.
Davis, S.L., Garrigus, U.S. and Hinds, F.C. 1970. Metabolic effects of
    growth hormone and diethylstilbestrol in lambs. II. Effects of
    daily ovine growth hormone injections on plasma metabolites and
    nitrogen-retention in fed lambs. J. Animal Sci. 30: 236-240.
Davis, S.L., Hill, K.M., Ohlson, D.L. and Jacobs, J.A. 1976. Influence of
    chronic thyrotropin-releasing hormone injections on secretion of
    prolactin, thyrotropin and growth hormone and on growth rate in
    wether lambs. J. Animal Sci. 42: 1244-1250.
Davis, S.L., Ohlson, D.L., Klindt, J. and Everson, D.O. 1979. Estimates
    of repeatability in the temporal patterns of secretion of growth
    hormone (GH), prolactin (PRL) and thyrotropin (TSH) in sheep. J.
    Animal Sci. 49: 724-728.
Davis, S.L., Ohlson, D.L., Klindt, J. and Anfinson, M.S. 1977a. Episodic
    growth hormone secretory patterns in sheep: Relationship to gonadal
    steroid hormones. Am. J. Physiol.: Endocrinol. Metab. Gastrointest.
    Physiol. 6: E519-E523.
Davis, S.L., Ohlson, D.L., Blann, D.L., Svoboda, M.E. and Van Wyk, J.J.
    1981. Purification and partial characterization of somatomedin
    activity from sheep serum. J. Animal Sci. 53: Suppl. 1, 304-305,
    Abstr.
Davis, S.L., Sasser, R.G., Thacker, D.L. and Ross, R.H. 1977b. Growth
    rate and secretion of pituitary hormones in relation to age and
    chronic treatment with thyrotropin-releasing hormone in prepubertal
    dairy heifers. Endocrinol. 100: 1394-1402.
de Bodo, R.C. and Altszuler, N. 1957. The metabolic effects of growth
    hormone and their physiological significance. Vit. Hormones 15:
    205-208.
Desjardins, C. 1981. Endocrine signaling and male reproduction. Biol.
    Reprod. 24: 1-21.
Dodson, M.V., Davis, S.L., Ohlson, D.L. and Ercanbrack, S.K. 1983.
    Temporal patterns of growth hormone, prolactin and thyrotropin
    secretion in Targhee rams selected for rate and efficiency of gain.
    J. Animal Sci. 57: In press.
Driver, P.M. and Forbes, J.M. 1980. Episodic growth hormone secretion in
    sheep in relation to time of feeding, spontaneous meals and short
    term fasting. J. Physiol. 317: 413-424.
Dulak, N.C. and Temin, H.M. 1973a. A partially purified polypeptide
    fraction from rat liver cell conditioned medium with multiplication-

stimulating activity for embryo fibroblasts. J. Cell. Physiol. 81: 153-160.

Dulak, N.C. and Temin, H.M. 1973b. Multiplication-stimulating activity for chicken embryo fibroblasts from rat liver cell conditioned medium: A family of small polypeptides. J. Cell. Physiol. 81: 161-170.

Eberhardt, N.L., Apriletti, J.W. and Baxter, J.D. 1980. The molecular biology of thyroid hormone action. In "Biochemical Actions of Hormones" (Ed. G. Litwack). (Academic Press, New York). Vol. VII. pp. 311-394.

Ewton, D.Z. and Florini, J.R. 1980. Relative effects of the somatomedins, multiplication-stimulting activity and growth hormone on myoblasts and myotubes in culture. Endocrinol. 106: 577-583.

Ewton, D.Z. and Florini, J.R. 1981. Effects of somatomedins and insulin on myoblast differentiation. Develop. Biol. 86: 31-39.

Fain, J.N. 1962. Effects of dexamethasone and growth hormone on fatty acid mobilization and glucose utilization in adrenalectomized rats. Endocrinol. 71: 633-635.

Fain, J.N. 1968. Effect of dibutyryl-3',5'-AMP, theophylline and norepinephrine on lipolytic action of growth hormone and glucocorticoid in white fat cells. Endocrinol. 82: 825-830.

Falconer, J., Buttle, H.J. and Forbes, J.M. 1977. Somatomedin levels in the blood of hypophysectomized sheep given growth hormone or prolactin. J. Endocrinol. 72: 30P.

Falconer, J., Forbes, J.M., Bines, J.H., Roy, J.H.B. and Hart, I.C. 1980. Somatomedin-like activity in cattle: The effect of breed, lactation and time of day. J. Endocrinol. 86: 183-188.

Finkelstein, J.W., Roffwarg, H.P., Boyar, R.M., Kream, J. and Hellman, L. 1972. Age-related change in the twenty-four-hour spontaneous secretion of growth hormone. J. Clin. Endocrinol. Metab. 35: 665-670.

Florini, J.R. 1983. Hormonal control of muscle cell growth. In "Current Concepts of Animal Growth" (Ed. C.E. Allen). (Am. Soc. Animal Sci., Champaign, Ill.) In press.

Francis, M.J.O. and Hill, D.F. 1975. Prolactin stimulated production of somatomedin by rat liver. Nature 255: 167-168.

Froesch, E.R., Burgi, J., Ramseier, E.B., Bally, P. and Labhart, A. 1963. Antibody-suppressible and nonsuppressible insulin-like activities in human serum and their physiological significance. An insulin assay with adipose tissue of increased precision and specificity. J. Clin. Invest. 42: 1816-1834.

Galbraith, H., Dempster, D.G. and Miller, T.B. 1978. A note on the effect of castration on the growth performance and concentrations of some blood metabolites and hormones in British Friesian male cattle. Anim. Prod. 26: 339-342.

Galbraith, H. and Geraghty, K.J. 1982. A note on the response of British Friesian steers to trenbolone acetate and hexestrol, and to alteration in dietary feed intake. Anim. Prod. 35: In press.

Giordano, G., Van Wyk, J.J. and Minuto, F. (Eds.). 1979. "Somatomedins and Growth." (Academic Press, New York). 363 pp.

Goldberg, A.L., Tischler, M., DeMartino, G. and Griffin, G. 1979. Hormonal regulation of protein degradation. Fed. Proc. 39: 31-36.

Gospodarowicz, D., Wesemen, J.J., Moran, S. and Lindstrom, J. 1976. Effect of fibroblast growth factor on the division and fusion of bovine myoblasts. J. Cell. Biol. 70: 395-405.

174

Guillemin, R., Brazeau, P., Bohlen, P., Esch, F., Ling, N. and Wehren-
     berg, W.B. 1982. Growth hormone-releasing factor from a human
     pancreatic tumor that caused acromegaly. Science 218: 585-587.
Hall, K. 1972. Human somatomedin. Determination, occurrence, biological
     activity and purification. Acta Endocrinol. 70 Supp. 163, 5-52.
Heitzman, R.J. 1981. Mode of action of anabolic agents. In "Hormones and
     Metabolism in Ruminants" (Ed. J.M. Forbes and M.A. Lomax). (Agricul-
     tural Research Council, London). pp. 129-139.
Hintz, R.L. and Liu, F. 1981. Insulin-like growth factor. II. Radioim-
     munoassay based on an antiserum against the synthetic C-peptide
     fragment. Clin. Res. 29: 409A.
Holder, A.T., Spencer, E.M. and Preece, M.A. 1981. Effect of bovine
     growth hormone and a partially purified preparation of somatomedins
     on various growth parameters in hypopituitary dwarf mice. J.
     Endocrinol. 89: 275-282.
Hove, K. and Blom, A.K. 1973. Plasma insulin and growth hormone in dairy
     cows: Duirnal variation and relation to food intake and plasma sugar
     and acetoacetate levels. Acta Endocrinol. 73: 289-303.
Jansson, J.-O., Albertsson-Wikland, K., Eden, S. Thorngren, K.-G. and
     Isaksson, O. 1982a. Effect of frequency of growth hormone adminis-
     tration on longitudinal bone growth and body weight in hypophysec-
     tomized rats. Acta Physiol. Scand. 114: 261-265.
Jansson, J.-O., Albertsson-Wikland, K., Eden, S., Thorngren, K.-G. and
     Isaksson, O. 1982b. Circumstantial evidence for a role of the
     secretory pattern of growth hormone in control of body growth. Acta
     Endocrinol. 99: 24-30.
Keller, D.G., Smith, V.G., Coulter, G.H. and King, G.J. 1979. Serum
     growth hormone concentration in Hereford and Angus calves: Effects
     of breed, sire, sex, age, age of dam and diet. Can. J. Animal Sci.
     59: 367-373.
Kertz, A.F., Reid, J.T., Wellington, G.H., Ayala, H.J., Simpfendorfer,
     S., Maiga, A.M. and Fortin, A. 1982. Growth and development of
     cattle as related to breed, sex and level of intake. Search:
     Agriculture. Ithaca, NY: Cornell Univ. Agr. Exp. Sta. No. 23, 32
     pp.
Knobil, E. 1981. Patterns of hypophysiotropic signals and gonadotropin
     secretion in the rhesus monkey. Biol. Reprod. 24: 44-49.
Knobil, E. and Hotchkiss, J. 1964. Growth hormone. Ann. Rev. Physiol.
     26: 47-74.
Kostyo, J.L. and Knobil, E. 1959. The effect of growth hormone on the in
     vitro incorporation of leucine-2-C14 into the protein of rat
     diaphragm. Endocrinol. 65: 395-401.
Kuhlemeier, K.V. and Trenkel, A. 1966. Effects of growth hormone on lipid
     metabolism in ruminants. J. Animal Sci. 25: 1265-1266.
Li, C.H. and Evans, H.M. 1944. The isolation of pituitary growth hormone.
     Science 99: 183-184.
Lund-Larsen, T.R., Sundby, A., Kruse, V. and Velle, W. 1977. Relation
     between growth rate, serum somatomedin and plasma testosterone in
     young bulls. J. Animal Sci. 44: 189-194.
Machlin, L.K. 1972. Effect of porcine growth hormone on growth and
     carcass composition of the pig. J. Animal Sci. 35: 794-800.
Manchester, K., Randle, P.J. and Young, F.G. 1959. The effect of growth
     hormone and of cortisol on the response of isolated rat diaphragm to
     the stimulating effect of insulin on glucose uptake and on incor-
     poration of amino acids and protein. J. Endocrinol. 18: 395-408.

175

Martin, J.B. 1974. Regulation of the pituitary-thyroid axis. In "MTP International Review of Science: Endocrine physiology" (Ed. S.M. McCann). (Butterworths, London and University Park Press, Baltimore). Vol. 5: 96-97.

Marx, J.L. 1982. Still more about gene transfer. Science 218: 459-460.

McAtee, J.W. and Trenkle, A. 1971. Effect of feeding, fasting and infusion of energy substrates on plasma growth hormone levels in cattle. J. Animal Sci. 33: 612-616.

McConaghey, P.D. and Sledge, C.B. 1970. Production of "sulphation factor" by the perfused liver. Nature 255: 1249-1250.

McGuffey, R.K., Thomas, J.W. and Convey, E.M. 1977. Growth, serum growth hormone, thyroxine, prolactin and insulin in calves after thyrotropin releasing hormone or 3-methyl-thyrotropin releasing hormone. J. Animal Sci. 44: 422-430.

Merimee, T.J. and Rabin, ·D. 1973. A survey of growth hormone secretion and action. Metabolism 22: 1235-1251.

Merriam, G.R. and Wachter, K.W. 1982. Algorithms for the study of episodic hormone secretion. Am. J. Physiol.: Endocrinol. Metab. 243: E310-E318.

Michel, G. and Baulieu, E.-E. 1980. Androgen receptor in rat skeletal muscle: Characterization and physiological variations. Endocrinol. 107: 2088-2098.

Moseley, W.M., Krabill, L.F. and Olsen, R.F. 1982. Effect of bovine growth hormone (GH) administered in various patterns on nitrogen metabolism in the Holstein steer. J. Animal Sci. 55: 1062-1070.

Mosier, H.D., Jr. 1981. Thyroid hormone. In "Endocrine Control of Growth" (Ed. W.H. Daughaday). (Elsevier, New York). pp. 25-66.

Muir, L.A., Wien, S., Duquette, P.F., Rickes, E.L. and Cordes, E.H. 1983. Effects of exogenous growth hormone and diethylstilbestrol on growth and carcass composition of growing lambs. J. Animal Sci. 56: In press.

Noall, M.W., Riggs, T.R., Walker, L.M. and Christensen, H.N. 1957. Endocrine control of amino acid transfer. Distribution of an unmetabolizable amino acid. Science 126: 1002-1005.

Ohlson, D.L., Davis, S.L., Ferrell, C.L. and Jenkins, T.G. 1981. Plasma growth hormone, prolactin and thyrotropin secretory patterns in Hereford and Simmental calves. J. Animal Sci. 53: 371-375.

Olsen, R.F., Wangsness, P.J., Patton, W.H. and Martin, R.J. 1981. Relationship of serum somatomedin-like activity and fibroblast proliferative activity with age and body weight gain in sheep. J. Animal Sci. 52: 63-68.

Olson, D.P., Ohlson, D.L., Davis, S.L. and Laurence, K.A. 1981. Acidophil adenoma in the pituitary gland of a sheep. Vet. Pathol. 18: 132-135.

Pecile, A. and Muller, E.E. (Eds.) 1976. "Growth Hormone and Related Peptides." Proc. 3rd Int. Symp. Excerpta Medica Int. Congr. Series No. 381. 466 pp.

Phillips, L.S., Herington, A.C. and Daughaday, W.H. 1975. Steroid hormone effects on somatomedin. I. Somatomedin action in vitro. Endocrinol. 97: 780-786.

Phillips, L.S. and Vassilopoulou-Sellin, R. 1980a. Somatomedins, I. New Engl. J. Med. 302: 371-380.

Phillips, L.S. and Vassilopoulou-Sellin, R. 1980b. Somatomedins, II. New Engl. J. Med. 302: 438-446.

Pierson, R.W., Jr. and Temin, H.M. 1972. The partial purification from calf serum of a fraction with multiplication-stimulating activity for chicken fibroblasts in cell culture and with non-suppressible insulin-like activity. J. Cell. Physiol. 79: 319-329.

Reynert, R., Marcus, S., DePaepe, M. and Peeters, G. 1976. Influence of stress, age and sex on serum growth hormone and free fatty acid levels in cattle. Horm. Metab. Res. 8: 109-114.

Riggs, T.R. and Walker, L.M. 1960. Growth hormone stimulation of amino acid transport into rat tissues in vivo. J. Biol. Chem. 235: 3603-3607.

Rinderknecht, E. and Humbel, R.E. 1976. Amino terminal sequences of two polypeptides from human serum with nonsuppressible insulin-like and cell-growth-promoting activities: Evidence for structural homology with insulin B chain. Proc. Nat. Acad. Sci. USA, 73: 4379-4381.

Rinderknecht, E. and Humbel, R.E. 1978a. The amino acid sequence of human insulin-like growth factor I and its structural homology with proinsulin. J. Biol. Chem. 253: 2769-2776.

Rinderknecht, E. and Humbel, R.E. 1978b. Primary structure of human insulin-like growth factor II. FEBS Lett. 89: 283-286.

Ringberg, T. 1979. Serum somatomedin - a measure of prospective growth capacity in bull calves. Acta Agric. Scand. 29: 216-218.

Russell, J.A. 1957. Effects of growth hormone on protein and carbohydrate metabolism. Am. J. Clin. Nutr. 5: 404-416.

Saenger, P., Schwartz, E., Wiedemann, E., Levine, L.S., Tsai, M. and New, M.I. 1976. The interaction of growth hormone, somatomedin and oestrogen in patients with Turners syndrome. Acta Endocrinol. 81: 9-18.

Salmon, W.D., Jr. and Daughaday, W.H. 1957. A hormonally controlled serum factor which stimulates sulfate incorporation by cartilage in vitro. J. Lab. Clin. Med. 49: 825-836.

Salmon, W.D., Bower, P.H. and Thompson, E.Y. 1963. Effect of protein anabolic steroids on sulfate incorporation by cartilage of male rats. J. Lab. Clin. Med. 61: 120-128.

Samuels, H.H., Perlman, A.J., Raaka, B.M. and Stanley, F. 1982. Organization of the thyroid hormone receptor in chromatin. Rec. Prog. Hormone Res. 38: 557-592.

Santen, R.J. and Bardin, C.W. 1974. Episodic luteinizing hormone secretion in man. J. Clin. Invest. 52: 2617-2628.

Schalch, D.S., Heinrich, U.E., Draznin, B., Johnson, C.J. and Miller, L.L. 1979. Role of the liver in regulating somatomedin activity: hormonal effects on the synthesis and release of insulin-like growth factor and its carrier protein by the isolated perfused rat liver. Endocrinol. 104: 1143-1151.

Schoenle, E., Zapf, J., Humbel, R.E. and Froesch, E.R. 1982. Insulin-like growth factor I stimulates growth in hypophysectomized rats. Nature 296: 252-253.

Snochowski, M., Dahlberg, E. and Gustafsson, J.-A. 1980. Characterization and quantification of the androgen and glucocorticoid receptors in cytosol from rat skeletal muscle. Eur. J. Biochem. 111: 603-616.

Tannenbaum, G.S. and Martin, J.B. 1976. Evidence for an endogenous ultradian rhythm governing growth hormone secretion in the rat. Endocrinol. 98: 562-570.

Tindal, J.S., Knaggs, G.S., Hart, I.C. and Blake, L.A. 1978. Release of growth hormone in lactating and non-lactating goats in relation to behaviour, stages of sleep, electroencephalograms, environmental stimuli and levels of prolactin, insulin, glucose and free fatty acids in the circulation. J. Endocrinol. 76: 333-346.

Trenkle, A. 1970. Plasma levels of growth hormone, insulin and plasma protein-bound iodine in finishing cattle. J. Animal Sci. 31: 389-393.

Trenkle, A. 1976. Estimates of the kinetic parameters of growth hormone metabolism in fed and fasted calves and sheep. J. Animal Sci. 43: 1035-1043.

Trenkle, A. 1978. Relation of hormonal variations to nutritional studies and metabolism of ruminants. J. Dairy Sci. 61: 281-293.

Trenkle, A. and Topel, D.G. 1978. Relationships of some endocrine measurements to growth and carcass composition of cattle. J. Animal Sci. 46: 1604-1609.

Underwood, L.E., D'Ercole, A.J., Copeland, K.C., Van Wyk, J.J., Hurley, T. and Handwerger, S. 1982. Development of a heterologous radio-immunoassay for somatomedin-C in sheep blood. J. Endocrinol. 93: 31-39.

Uthne, K. 1973. Human somatomedins: Purification and some studies on their biological actions. Acta Endocrinol. 73: Suppl. 175, 1-35.

van Buul-Offers, S. and Van den Brande, L. 1979. Effect of growth hormone and peptide fractions containing somatomedin activity on growth and cartilage metabolism of snell dwarf mice. Acta Endocrinol. 92: 242-257.

van Buul-Offers, S. and Van den Brande, J.L. 1982. Cellular growth of dwarf mice during treatment with growth hormone, thyroxine and plasma fractions containing somatomedin activity. Acta Endocrinol. 99: 150-160.

Van Cauter, E., L'Hermite, M., Copinschi, G., Refetoff, S., Desir, D. and Robyn, C. 1981. Quantitative analysis of plasma prolactin in normal man. Am. J. Physiol.: Endocrinol. Metab. Gastrointest. Physiol. 241: E355-E363.

Van den Brande, J.L., Kootte, F., Tielenburg, R., van der Wilk, M. and Huyser, T. 1974. Studies on plasma somatomedin activity in different animal species. Acta Endocrinol. 75: 243-248.

Van Wyk, J.J., Underwood, L.E., Hintz, R.L., Clemmons, D.R., Voina, S.J. and Weaver, R.P. 1974. The somatomedins: A family of insulin-like hormones under growth hormone control. Rec. Prog. Horm. Res. 30: 259-295.

Van Wyk, J.J. and Underwood, L.E. 1978. The somatomedins and their actions. In "Biochemical Actions of Hormones" (Ed. G. Litwack). (Academic Press, N.Y.). Vol. 5, pp. 101-148.

Verde, L.S. and Trenkle, A. 1982. Concentration of hormones in plasma from cattle with different growth potentials. J. Animal Sci. 55: Supp. 1, 230, Abstr.

Wangsness, P.J., Olsen, R.F. and Martin, R.J. 1981. Effects of breed and zeranol implantation on serum insulin, somatomedin-like activity and fibroblast proliferative activity. J. Animal Sci. 52: 57-62.

Wheatley, I.S., Wallace, A.L.C. and Bassett, J.M. 1966. Metabolic effects of ovine growth hormone in sheep. J. Endocrinol. 35: 341-353.

Wiedemann, E. 1981. Adrenal and gonadal steroids. In "Endocrine Control of Growth" (Ed. W.H. Daughaday). (Elsevier, New York). pp. 67-119.

Wiedemann, E. and Schwartz, E. 1972. Suppression of growth hormone-dependent human serum sulfation factor by estrogen. J. Clin. Endocrinol. Metab. 34: 51-58.

Wiedemann, E., Schwartz, E. and Frantz, A.G. 1976. Acute and chronic estrogen effects upon serum somatomedin activity, growth hormone and prolactin in man. J. Clin. Endocrinol. Metab. 42: 942-952.

Williams, I.H., Chua, B.H.L., Sahms, R.H., Siehl, D. and Morgan, H.E. 1980. Effects of diabetes on protein turnover in cardiac muscle. Am. J. Physiol.: Endocrinol. Metab. Gastrointest. Physiol. 239: E178-E185.

Wool, I.G., Stirewalt, W.S., Kurihara, K., Low, R.B., Bailey, P. and Oyer, D. 1968. Mode of action of insulin in the regulation of protein biosynthesis in muscle. Rec. Prog. Hormone Res. 24: 139-213.

Young, F.G. 1953. The growth hormone and diabetes. Rec. Prog. Hormone Res. 8: 471-510.

Zapf, J., Walter, H. and Froesch, E.R. 1981. Radioimmunological determination of insulin-like growth factors I and II in normal subjects and patients with growth disorders and extrapancreatic tumor hypoglycemia. J. Clin. Invest. 68: 1321-1330.

DISCUSSION ON DR. DAVIS'S PAPER

Prof. Karg: Do growth hormone and somatomedins influence or control each other?

Dr. Davis: Yes. Reports in the literature indicate that treatment with somatomedin does inhibit growth hormone secretion. There may be a feedback relationship between growth hormone and somatomedin much similar to thyroid hormone feedback on TSH release.

Dr. Forbes: I would place a different emphasis than you on the relationship of feeding, whether spontaneous or restricted, on growth hormone secretion. This makes interpretation of results rather difficult. You did not say, except in one case, what the level or frequency of feeding was in your experiments. You can imagine a situation where animals are fed the same amount of food but one group has a higher growth ptoential and therefore will be relatively more restricted in terms of nutrients. Because of the well known effect of lack of nutrients on growth hormone secretion, this will give higher growth hormone levels. The interpretation that faster growing animals have higher growth hormone levels is not direct evidence of animals with higher growth potential having higher growth hormone.

Dr. Davis: That is a possible interpretation of the studies conducted. We would like to establish whether or not growth hormone secretory patterns are indeed related to different growth rates. In other words if we were to select animals for their ability to secrete growth hormone would that kind of selection result in an increased rate of growth?

Dr. Forbes: Not necessarily - from the arguements I have just given. It depends on so many things. You should put more emphasis on the type and level of feeding. Most of your experiments have been done with fasting animals.

Dr. Davis: Not necessarily. The studies with cattle of different breeds and sheep of different selected lines were done with fed animals.

Prof. Karg: How long did you let the animals fast? For how long did you see the increase in growth hormone which you explained by perhaps a slower metabolism and clearance? In our studies with prolactin in the bovine, we had up to 2 - 3 days of fasting which always gave a decrease. Could you tell me the time curves of growth hormone during fasting?

Dr. Davis: In the studies I was referring to, which were not our studies, probably anywhere from 2 days to a week. The magnitude of the growth hormone response to the fasting is generally speaking quite small. Possibly that is why there has been disagreement in the literature as to whether or not fasting in all cases increases plasma growth hormone concentrations.

Dr. Galbraith: Are you confident that what you are measuring is in fact increases in secretion and not variations in the breakdown of the growth hormone? Is it possible you are picking up breakdown products in the radioimmunoassay which are not biologically active?

Dr. Davis: The possibility that we are measuring metabolic products which may or may not be associated with growth is a very real possibility. I have no data to prove that the R.I.A. which we are using, measures something of equivalent biologic activity. This is the same disadvantage that all radioimmunoassays possess because you are measuring immunological activities viz. activity associated with the antigenic site which is a very small portion of the molecule. That could mean that a substantial amount of metabolic degradation products are being measured. However the fact that we are seeing changes in what we are measuring which are associated with growth, may suggest that what we are measuring is related to biological activity.

Dr. Sejrsen: You put emphasis on the secretory pattern
but there could be a difference between lactation and growth.
In experiments using exogenous growth hormone the lactation
response from similar amounts of growth hormone given either
every other day, every six hours or by continuous infusion have
been the same.

Dr. Davis: I am familiar with those data. Dr. Baumann at
Cornell has observed that he gets the same lactation
response whether he administered growth hormone by multiple
daily injections or by continuous infusion. However, in
that study, only one dose and multiple injection schedule
was used. It is possible that injection of different doses
at different frequencies or at different times of
day may yet prove to produce a greater biological`response than
continuous infusion.

Dr. Sejrsen: Could it be related to the different levels
of endogenous GH?

Dr. Davis: It is very possible.

Dr. van der Waal: You suggested injection of growth hormone
or somatomedin as possible ways to increase growth. Are you
aware of any developments of adequate releasing mechanisms
that can be used to do this in farm animals?

Dr. Davis: No, none at all.

Dr. Clancy: You referred to the hormonal influence on
muscle cell division, the view generally accepted is that
the number of muscle fibres are fixed at birth in farm
animals. The effect you referred to must take place in
the foetus, and go back to the presumptive myoblast stage
where you still get cellular division. After a while the
myoblasts loose their ability to divide and they fuse into

182

myotubes and so on. I take it you are referring to that
early effect and it leads me to wonder, whether or not
there is a maternal transplacental effect operating on the
foetus.

Dr. Davis: What you are saying looks very possible.
In fact studies which were done with somatomedins were done
with presumptive myoblasts and it was shown that somatomedins
could stimulate both division or mitosis of the presumptive
myoblasts and also fusion to form new myotubes. There
is however a recent suggestion within the last few years that
cells in post-natal muscle may also be capable of dividing
in response to certain stimuli such as somatomedins.
The cells which I am referring to specifically are the
satellite cells. Possibly one of the means by which post-natal
growth occurs is by the addition of more nuclei to existing
muscle fibres. You are correct when you say it appears
that the number of muscle fibres may be fixed at birth, but
the size of that fibre may increase by protein synthesis within
that fibre and by the addition of more nuclei or satellite
cells to existing muscle fibres.

Dr. Clancy: This is important for future work in relation
to meat development. Unlocking of the function of
satellite cells is critical and I believe they are in
G 1 arrest.

Dr. Heitzman: Dr. Spencer in his paper made the categoric
claim that sex steroids do not affect bone length,
but Dr. Robertson and yourself both suggest that alterations
in sex steroid concentration may affect bone length, or
skeletal development, by virtue of their effect on growth
hormone or somatomedins. Would you comment?

Dr. Davis: In my opinion there may be two effects of the
gonadal steroids. One is to stimulate long bone growth directly.
When you see the change in growth, and height of young girls
at puberty it appears rather obvious that there is something

which is triggering that change in growth and bone length.
Most of the information suggests that this is a gondal steroid
effect. Secondly as suggested in the literature that
oestrogens and androgens may be different in their ability
to ultimately cause closure of the epiphysis, in other
words to actually stop or arrest growth in height. The
oestrogens are more effective there than the androgens,
and perhaps this is one of the reasons why females generally
stop growing in height at a level considerably less than in
males. Thus the different balance of oestrogens and androgens
produced by the different sexes may explain differences in
bone growth.

Dr. Schanbacher: Would you comment on the possible temporal
relationship between growth hormone and somatomedin if it
is known in any species. Perhaps either you or Dr. Spencer
could comment on thoughts of what might be going on in
lambs immunized against somatostatin.

Dr. Davis: I'm not sure I want to comment on the latter.
I believe there is limited information on the temporal
relationship between growth hormone and somatomedin
secretion. Studies in humans suggest that, although
growth hormone is secreted episodically somatomedin secretion
is quite constant over time, it is not secreted in an episodic
fashion.

Dr. Forbes: We have seen fairly constant somatomedin levels
in sheep and cattle. However it does go up and down in
a way that suggests the half-life is less than has been
reported in the human - say 12 hours.

Dr. Davis: At least it is more stable in blood than growth
hormone.

MODES OF ACTION IN MANIPULATING RUMEN FUNCTION

Gerald T. Schelling

Department of Animal Science
Texas A&M University
College Station, Texas 77843, USA

ABSTRACT

The ionophore monensin is used as a model to examine the modes of action that are important in manipulating rumen function. It appears likely that several system modes of action come into play and that they result from the basic mode of action of the ionophore which involves the modification of the movement of ions across the biological membranes of rumen microbes. While there are a number of biological claims reported in the literature for monensin, they can be consolidated to seven categories or system modes of action. The modification of acid production is one well recognized category of great importance. Modified feed intake should also be considered to be important. The third system mode of action, change in gas production, probably only contributes a limited savings in energy. Modified digestibilities are probably quite variable as a mode of action, but would appear to be a significant factor. The change in protein utilization appears to result from several factors that are probably occuring simultaneously. Modification of rumen fill and rate of passage may play an important role in causing some of the previously mentioned system modes of action to occur, but this is not presently clear. A seventh category inclusive of several monensin responses that are more indirect to the rumen or sporadic in nature is included. It appears that monensin responses occur as a result of these several system modes of action which probably act in concert. It is impossible to accurately assess a quantitative contribution of each of these categories at the present time.

INTRODUCTION

The rumen microbes provide unique capabilites to the ruminant in regard to feed utilization. While the fermentation in the rumen offers definite advantages to the host animal, it has long been the desire of researchers to gain control of the metabolism. Being able to modulate pathways involving energy and nitrogen metabolism would yield control that could be of considerable economic importance. It has only been within the past ten years that chemical agents have been identified to offer significant potential for the manipulation of rumen function.

It is the intent of this paper to cite some of the more prominent compounds which have been identified as having the potential to intervene in rumen metabolism, and then examine in more detail the modes of action that appear to be involved. The compound monensin provides an excellent tool to study these modes of action due to its intensive investigation in recent years and its demonstrated widespread effectiveness to enhance animal production.

CHEMICAL ADDITIVES TO THE RUMEN

It has been fairly common to classify the chemical additives as propionate enhancers and methane, deaminase and lactic acid inhibitors. While this is a convenient framework to think of, more recent work has demonstrated that most of the chemical agents fit into several of these classes. Thus it is more realistic to consider each compound on the basis of its individual characteristics.

The ionophores constitute a group of compounds enjoying considerable success as feed additives. In addition to monensin, lasalocid is now being marketed for use as a ruminant feed additive. Numerous references to monensin will be made subsequently in this paper. The reported studies with lasalocid have been consolidated (Stuart and Zimmerman, 1982), and have indicated similar responses to those of monensin. Salinomycin is another ionophore that results in improved beef cattle performance (McClure et al., 1980), and produces intraruminal volatile fatty acid changes similar to monensin

(Fontenot et al., 1980). Salinomycin has only been studied to a
limited extent. Narasin is yet another ionophore that appears to
mimic monensin (Potter et al., 1979). It has undergone moderate
investigation and appears to be more potent than monensin. Although
not an ionophore, avoparcin is a glycopeptide antibiotic that
produces performance responses in beef cattle (Ingle et al., 1978).
The work reported with avoparcin has been consolidated (Anonymous,
1982) and indicates that avoparcin also produces responses quite
similar to monensin.

A family of diaryliodonium compounds have been identified as
potential deaminase inhibitors (Chalupa, 1977), but have not been
extensively studied in recent years. Dimethyldiphenyliodonium
chloride has been the compound most frequently studied from this
laboratory. Halogenated methane analogs have been identified as
methane inhibitors in the rumen (Trei et al., 1970). Amicloral, a
hemiacetal of chloral hydrate and starch, has resulted in some
improvement in feed efficiency when fed to ruminants (Chalupa, 1980).
Thiopeptin, a sulfur-containing peptide antibiotic, has been
demonstrated to inhibit lactic acid production in the rumen (Muir et
al., 1981), and has also resulted in improved performance when fed to
cattle (Gill et al., 1979). A review of other antibiotics and
agricultural chemicals has been prepared (Chalupa, 1980).

While the ionophore monensin has been widely used for feedlot
cattle since 1975, most researchers do not feel that the mechanisms
whereby monensin produces increased performance are fully recognized.
While some of the effects of monensin are well established, and
perhaps even account for most of the performance response, there is

much work that suggests additional modes of action. While monensin has been the leading ionophore as a feed additive for cattle and will be used to study the modes of action in this paper, it is important to recognize that more than 70 ionophores have been identified to date. Much information concerning the utility of ionophores is currently available, and it is likely that the near future will yield other beneficial roles in relationship to nutrition and animal production.

The general basic mode of action of monensin and other ionophores to modify the movement of ions across biological membranes is considered to be the fundamental action involved. However, it is the intent of this paper to discuss the system modes of action which implement the animal performance response normally observed when monensin is fed to ruminants. It is assumed that animal responses result from the system modes of action which result directly from the basic mode of action.

TYPES OF MONENSIN MODES OF ACTION

Monensin has been a very heavily studied chemical in the area of ruminant nutrition. Few materials have undergone as extensive investigation and probably none have been implicated as producing as many biological changes. The volume of literature concerning monensin offers a formidable challenge to consolidate into likely modes of action.

It does appear that there is one type of basic mode of action for monensin at the cellular level. It involves the transport of ions across biological membranes. In fact, this basic function has yielded the name ionophore or "ion bearer". The effectiveness of monensin to cause sodium entry into cells has been well established (Smith and Rosengurt, 1978), and other reported cellular ion changes may be secondary to the primary sodium effect (Schanne et al., 1979). This basic cellular mode of action is well documented and has been reviewed (Pressman, 1976).

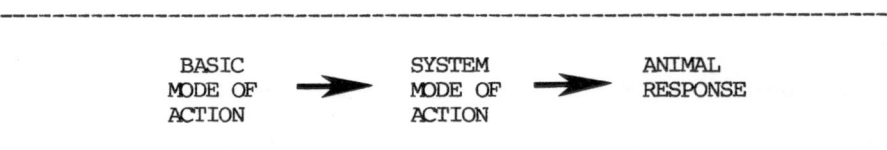

Figure 1   Mode of action relationships.

It is the implementation of system mode(s) of action which more directly cause animal responses that is of special interest to nutritionists. The early work with monensin and rumen microbes readily demonstrated an increased propionate to acetate ratio (Richardson et al., 1976), and this change is still recognized as a very important one. But as more work has accumulated with monensin it has become clear that other changes are occurring that are probably system modes of action in the rumen. The wide array of literature would suggest that the basic mode of action of cellular ion transport results in some important changes which influence the rumen microbes in ways that lead to increased animal performance. It is the intent to discuss only these established and potential system modes of action that occur in the rumen. The reader should be aware that there will be no attempt to discuss the potential tissue modes of action that might occur with monensin.

BIOLOGICAL CLAIMS FOR MONENSIN IN THE RUMEN

The observed animal performance improvements of improved feed
efficiency in the feedlot and improved rate of gain in pasture-fed
ruminants have lead to experiments indicating a host of biological
changes occurring with monensin. A listing of biological claims that
have been reported is extensive (Table 1). While there is literature
to contradict some of the claims listed in the table, experimental
designs will often explain apparent differences. The claims listed
are well documented, although all research is not necessarily cited,
and are felt to be legitimate by the author.

TABLE 1   Biological claims for monensin in the rumen.

Item	Reference
Greater ruminal propionate concentration	Richardson et al., 1976
Lower ruminal acetate concentration	Richardson et al., 1976
Lower ruminal butyrate concentration	Richardson et al., 1976
Lower ruminal lactate in stressed animals	Dennis et al., 1980
Higher ruminal pH in stressed animals	Dennis et al., 1980
Less ruminal methane production	Chalupa et al., 1980
Decreased intake of grain diets	Raun et al., 1976
Increased intake of forage diets	Pond and Ellis, 1981
Increased ruminal forage fill	Ellis and Delaney, 1982
Decreased ruminal rate of passage	Lemanger et al., 1978
Increased dry matter digestibility	Rust et al., 1978
Increased protein digestibility	Beede et al., 1980
Decreased ruminal deamination	Schelling et al., 1977
Decreased ruminal proteolysis	Hanson et al., 1979
Protein sparing effect	Dartt et al., 1978
Modified ruminal escape of protein	Poos et al., 1979
Modified ruminal escape of starch	Rust et al., 1978
Modified rumen microbial population	Richardson et al., 1978
Increased body glucose turnover	Van Maanen et al., 1978
Modified substrate gluconeogenesis	Beede et al., 1980
Reduced 3-methylindole production	Hammond and Carlson, 1980
Earlier puberty in heifers	McCartor et al., 1979
Reduced fly pupae in feces	Herald et al., 1982

190

The items listed in Table 1 represent potential system modes of
action that might influence animal performance, and in most cases
they are probably the result of the basic momensin mode of action.
Modified rumen microbial metabolism quite likely accounts for most of
the listed observations, although an influence of monensin directly
on the gastrointestinal tissue could be involved with some of the
observations associated with rumen fill and rate of passage.
However, all of the listed items are viewed as being associated
primarily with ruminal effects. Effects such as increased body
glucose turnover and earlier puberty in heifers may appear to be
quite remote to the rumen, but there is evidence that they may be the
result of modifications in rumen fermentation.

Many of the ruminal effects due to monensin are closely related
although they have been listed (Table 1) as separate entities. To
facilitate the discussion of these many items they can be
consolidated into a limited number of groups on the basis of probable
relationships. This consolidation into accepted or probable system
modes of action of monensin are listed (Table 2) and will be
discussed in detail.

TABLE 2   Accepted or probable system modes of action of monensin.
----

        1. Modification of acid production.
        2. Modified feed intake.
        3. Change in gas production.
        4. Modified digestibilities.
        5. Change in protein utilization.
        6. Modified rumen fill and rate of passage.
        7. Other ruminal modes of action.

SYSTEM MODES OF ACTION OF MONENSIN

Modification of acid production

A highly documented effect of monensin in the rumen is its influence of decreasing the acetate to propionate ratio (Richardson et al., 1976). This ratio shift has long been viewed as a favorable change for meat producing ruminants. In fact several large industrial organizations have expended great effort to screen for compounds which would produce this effect, and monensin is the result of these efforts. This change in ratio presumes that there is actually more propionate produced at the expense of acetate, and has been documented to be the case with monensin through isotope dilution studies (Van Maanen et al., 1978). Ruminal butyrate concentrations are also normally decreased with monensin (Richardson et al., 1976).

The concept that propionate is more efficiently utilized than acetate or acetate precursors is based on two factors. The first is that propionate production by rumen fermentation appears to be more efficient than that of acetate. This has been discussed (Hungate, 1966) and reviewed more recently (Chalupa, 1977). The second factor, which is more controversial, is that there is evidence of propionate being utilized by the tissue more efficiently than acetate (Smith, 1971). Another possible advantage of propionate is that it is more flexible as an energy source than acetate. Propionate enjoys the luxury of having the potential to be used for gluconeogenesis in addition to direct oxidation by the citric acid cycle. Having more substrate for glycolysis may provide significant energetic advantages to the ruminant at certain times by generating more reduced coenzyme outside the mitochondrial membrane.

The monensin effect on volatile fatty acid production changes is well documented and widely accepted as one of the main system modes of action. However, it appears unlikely that this effect could account for all of the animal performance response normally observed with monensin. Studies (Raun et al., 1976) have suggested that improved ruminal energetic efficiency did not account for all of the feed efficiency response obtained with cattle. The increased fermentation efficiency response reported in some work (Owens, 1980) could account for a benefit equivalent only to a 1.6% increase in digestible energy. This conservation of energy would explain only a portion of the feed efficiency response with feedlot cattle. Even with the additional potential for some energy conservation at the tissue level due to propionate utilization, it would appear that other modes of action must be playing roles.

Monensin appears to cause a slight increase in lactate concentration (Beede and Farlin, 1977; Bartley et al., 1979) and no pH change (Dinius et al., 1976) in rumen fluid when cattle are not stressed with carbohydrate. However, with carbohydrate stress (Dennis et al., 1980) monensin produced a significant increase in pH and lower lactate concentration with in vitro studies of rumen fluid. Similar beneficial responses in cattle which were exposed to an induced lactic acidosis have been reported (Nagaraja et al., 1982). While these observations probably have little to do with the feed efficiency response observed with monensin-fed cattle, the work indicates some advantage for monensin to aid in the prevention of lactic acidosis.

The presumed basic mode of action role that monensin plays with the rumen microbes has not been elucidated. However, studies have indicated some definite changes in rumen microbial populations and/or microbial metabolism in response to monensin. Rumen protozoal numbers in cattle and sheep under pasture and several different feedlot conditions have been studied. The work indicates that monensin decreased protozoal numbers from 4 to 63% (Richardson et al., 1978). However, monensin may not decrease protozoal numbers under all conditions. No significant changes in numbers due to monensin with cattle fed a grass hay diet have been reported (Dinius et al., 1976).

The effects of monensin on metabolism by specific strains of rumen bacteria have been reported. Studies of types of rumen bacteria that might influence rumen volatile fatty acid ratios have been reported. Monensin has been found to be a metabolic inhibitor of hydrogen-producing and formate-producing bacteria, but was stimulatory for succinate-producing and propionate-producing bacteria (Chen and Wolin, 1978). An inhibition of hydrogen production by rumen microbes exposed to monensin has been reported (Van Nevel and Demeyer, 1977). The net effect of these influences would be to enhance propionate production at the expense of acetate production. The metabolism of hexose to lactate and subsequently to propionate has been studied relative to lactic acidosis (Dennis et al., 1980). It was found that monensin was inhibitory to four of seven strains of lactic acid producing rumen bacteria. This study also indicated that monensin did not influence the metabolism of three strains of rumen bacteria that ferment lactate to propionate, thus suggesting that monensin would tend to cause a decrease in lactate production from hexose and still maintain a normal capacity to metabolize any lactate produced to propionate.

Modified feed intake

It is well documented that monensin influences feed intake. Whether this phenomenon should be considered to be a system mode of action rather than classifying it as a response is debatable. However, for purposes of this discussion it will be considered as a system mode of action since feed intake is so important in the ruminant digestive physiology and absorption processes.

When cattle are under confinement conditions the feed intake depression due to to monensin is considerably greater for high grain diets than when high roughage diets are fed. The decreased feed intake with high grain diets has been reported to average 10.7% over a wide range of monensin feeding conditions (Anonymous, 1975b). However,this average includes severe feed depressions from abruptly introducing cattle to monensin which is likely to reduce feed intake by as much as 16%. Studies indicate depressions of less than 5% after 112 days on feed (Anonymous, 1975a). It would appear that realistic feed intake depressions with cattle fed high grain diets under desirable management conditions are about 5% (Owens, 1980) to 6% (Anonymous, 1975a). The feed intake depression with grain diets is definitely greater than that with roughage diets under confinement conditions. While more variable, the depression with higher roughage diets appears to be about 3% (Anonymous, 1975b). The increased rates of gains due to monensin with these types of diets are slight, but in conjunction with the decreased feed consumption the feed efficiency responses are significant.

Cattle receiving monensin under pasture conditions have responded with very significant increases in rate of gain and a summary of the work (Anonymous, 1975a) indicates a 17% rate of gain response. This response is difficult to interpret due to the inability to measure feed intake with standard grazing experiments. More recent studies using marker techniques to estimate the forage intake of grazing cattle fed monensin (Pond and Ellis, 1981) have suggested that cattle may be consuming up to 15% more forage to achieve the higher gains observed under pasture conditions. However, this work also suggests that the increased forage intake response may be related to the forage digestibility. It appears that the maximal 15% intake response due to monensin occurs at about 55% dry matter digestibility and the intake response diminishes to zero as digestibility increases to 65% or decreases to 45% (W.C. Ellis, unpublished data).

Changes in gas production

Another system mode of action for monensin is depressed methane production by rumen microbes. In vitro studies have indicated less microbial methane production with monensin (Bartley et al., 1979; Chalupa et al., 1980), and this response has also been demonstrated under in vivo conditions (Joyner et al., 1979; Thornton and Owens, 1981).

While essentially all of the reported studies indicate that monensin does reduce methane production, it is only a partial inhibition. The studies cited above resulted in reductions of 13 to 31% (Table 3). Other chemical agents can nearly completely inhibit methane production. It has been reported that amicloral, a halomethane analog, can inhibit in excess of 90% of the methane produced by rumen microbes (Trei et al., 1970). However, this methane inhibitor does result in some increase in hydrogen gas production whereas monensin does not (Chalupa et al., 1980). The effect of monensin on carbon dioxide production by rumen microbes appears to be negligible with lower levels of monensin, but is a significant depression at higher levels (Chalupa et al., 1980; Bartley et al., 1979).

TABLE 3. Monensin effect on methane productions.

% Change	Method	Reference
-21	in vitro	Bartley et al., 1979
-13	in vitro	Chalupa et al., 1980
-31	in vivo	Joyner et al., 1979
-16 to -24	in vivo	Thornton and Owens, 1981

Monensin has been reported (Chen and Wolin, 1978) to select against hydrogen-producing rumen bacteria and select for succinate-forming bacteria and Selenomonas ruminantium, which decarboxylates succinate to form propionate. These tendencies would lead to a decrease in methane formation. Other work (Van Nevel and Demeyer, 1977) has indicated that monensin causes a decrease in the metabolism of formate to carbon dioxide and hydrogen, and thus a decrease in methane production. Whatever the mechanism is, it appears that monensin causes a consistent, but low magnitude, depression of methane production. The low magnitude of the response probably contributes only a slight benefit to the enhanced efficiency of animal production.

Modified digestibilities

Several reports indicate an influence of monensin on digestibility, and thus suggests this as a system mode of action. The results are variable and the response is often not positive during adaptation periods. While the focus is primarily on extent of digestion, it may be that an effect on site of digestion and absorption may also play an important role. This might be especially important in the case of protein utilization, and this phenomenon will be discussed in that section.

The effect of adaptation to monensin on digestibility has not been thoroughly studied in an organized manner, but several studies suggest that monensin may initially produce a negative effect followed by a slight positive effect. It was found (Simpson, 1978) that monensin decreased cellulose digestibility when no adaptation time was allowed, but also found that monensin had no effect on cellulose digestibility when animals had been adapted to monensin for 21 days (Dinius et al., 1976). A reduction in dry matter and acid detergent fiber digestibility was observed when lambs were fed monensin for only 10 days, but after an additional 29 days of adaptation there was no significant difference (Poos et al., 1979).

Increases in dry matter and gross energy digestibilities have been reported in cattle fed a grain-roughage diet and adapted to monensin (Beede et al., 1980a). Other workers (Rust et al., 1978) reported an increased dry matter and starch digestibility in cattle fed monensin receiving a high grain diet which was low in protein. Marker techniques were used to measure an increased dry matter digestibility when cattle grazing bermudagrass were fed monensin (Pond and Ellis, 1981). Several studies indicate slight, but nonsignificant, increases in dry matter digestibility due to monensin (Thornton and Owens, 1981; Dinius et al., 1976; Poos et al., 1979). Neutral detergent fiber or acid detergent fiber digestibilities were not improved by monensin (Dinius et al., 1976; Beede et al., 1980a).

A variety of nitrogen digestibility trials have been reported concerning monensin. It is likely that the nature of the diet and the protein level in the diet are important factors. Increases in nitrogen digestibility due to monensin have been reported for cattle fed a low protein diet (Beede et al., 1980a; Rust et al., 1978), sheep fed a high protein diet (Joyner et al., 1979), and goats fed a low protein diet (Beede et al., 1978). Greater, but nonsignificant, nitrogen digestibilities were observed when cattle were fed low, medium or high roughage level diets (Thornton and Owens, 1981). Others have reported increased, but nonsignificant, responses (Dinius et al., 1976) or no difference (Poos et al., 1979) when urea containing diets were fed.

In general, it appears that monensin does result in a slight to moderate improvement in digestibility under many conditions. The conditions are not entirely clear at the present time, and it may be that factors such as level of feed intake, rumen fill and rumen rate of passage are important factors involved.

Change in protein utilization

Several cattle performance studies have suggested monensin may reduce the dietary protein requirement (Dartt et al., 1978; McCarthy et al., 1979). Thus there has been considerable interest in the effect of monensin on intraruminal nitrogen metabolism.

A variety of in vitro and in vivo studies have indicated that monensin significantly reduces the ruminal degradation of dietary protein (Schelling et al., 1977; Van Nevel and Demeyer, 1977; Chalupa, 1980; Poos et al., 1979). It has been demonstrated that monensin decreased the rate of free amino acid degradation in rumen fluid (Schelling et al., 1977). Rumen ammonia level decreases (Dinius et al., 1976) are consistent with the depression of deamination or proteolysis, or both.

The influence of monensin on rumen microbial protein synthesis appears to depend upon adaptation. Several studies have indicated that monensin decreases microbial growth when the microbes are not adapted to monensin (Bartley et al., 1979; Herod et al., 1979; Van Nevel and Demeyer, 1977). However, microbial growth by adapted cultures seems to be unaffected by monensin (Herod et al., 1979; Short et al., 1978). It has been observed that monensin decreased bacterial nitrogen reaching the abomasum of steers adapted to monensin (Poos et al., 1979). They also noted an increase in the amount of dietary protein reaching the abomasum and it is possible that the increased escape of dietary protein was of such a magnitude to limit available ammonia for bacterial growth. Others have obtained results that support a protein-sparing effect of monensin by encouraging ruminal escape of dietary protein (Short et al., 1978).

200

Several studies (Owens et al., 1978; Poos et al., 1979) have demonstrated that monensin increased the total amount of amino acid nitrogen reaching the abomasum. Work (Beede et al., 1980a) has demonstrated greater nitrogen retention due to feeding monensin to steers receiving a protein deficient diet, and monensin caused a greater animal performance response with steers fed low protein diets (Hanson and Klopfenstein, 1979). The review of literature strongly suggests that monensin does exhibit a protein-sparing effect by making more effective use of amino nitrogen.

Modified rumen fill and rate of passage

Rumen fill and rate of passage play a major role in ruminant nutrition. These factors influence the extent and site of digestibility, the extent of microbial fermentation, the end-products of fermentation and nitrogen utilization. It is beyond the scope of this paper to review this total area and several good reviews on the topic are available to the reader (Bull et al, 1979; Baldwin et al, 1977).

TABLE 4. Effect of monensin on rumen turnover rate and rumen fill.

Conditions	Phase	Rumen turn-over rate	Rumen fill
		---- % change ----	
Cattle grazing pasture[a]	solid	- 7	9
Cattle grazing range grass[b]	solid	-44	10
Cattle grazing range grass[b]	liquid	-31	—[c]
Cattle grazing range grass[b]	liquid	-22	24

[a]W.C. Ellis and K.R. Pond, unpublished data for 13 trials.
[b]Lemenager et al., 1978
[c]Not measured

In concentrating on discussing modified rumen fill and rate of passage as a system mode of action for monensin it readily becomes clear that many complexities will prohibit a full understanding at this point in time. However, the work conducted at a limited number of laboratories does consistently indicate that monensin decreases rumen turnover rate and increases rumen fill (Table 4). Although this work was conducted at only two different laboratories, it does represent 15 separate trials (Lemenger et al., 1978; Pond and Ellis, 1981; Ellis and Delaney, 1982).

While the work to this point indicates a monensin effect on rumen turnover rate and rumen fill, it is somewhat difficult to assess the implications of these changes. Several studies on the artificial induction of an increased rumen turnover rate have been reported (Harrison et al, 1975; Thomson et al, 1978). This work (Table 5) indicates that as liquid turnover rate was increased, the proportion of acetate produced by the microbes increased. Methane production and cell yield also increased with increased turnover rate. These changes resulted in a decreased fermentation efficiency. It is of considerable interest to note that the monensin ruminal responses discussed throughout this paper mimic the responses that result when rumen turnover decreases (Table 5). In fact, the similarity of the magnitude of the responses is so amazing that one is tempted to ascribe the monensin responses to the rumen fill and turnover effect. However, there is not yet adequate data to firmly establish this relationship. The importance of rumen turnover in ruminant nutrition is such that more research in this area is highly warranted.

202

TABLE 5. Effect of turnover rate on rumen fermentation in sheep.

MEASUREMENT (Moles/day)	Fraction liquid turnover rate per hour	
	.038	.098
Acetate	3.88	3.95
Propionate	1.83	1.23
Butyrate	.35	.62
Methane	5.76	6.88
Hexose fermented	3.47	3.32
Cell yield (g)	143.00	169.00
Fermentation efficiency (5)	77.10	73.00

Harrison et al., J. Agric. Sci. 85:93-101.

Other ruminal modes of action

Some of the biological claims reported for monensin (Table 1) do not fit into any of the previously discussed categories. Research has indicated that bovine pulmonary edema and emphysema can be experimentally produced by intraruminal tryptophan administration and the subsequent 3-methylindole production by rumen microbes. Studies have indicated that monensin decreases amino acid deamination in general (Schelling et al., 1977), and more specifically that it decreases 3-methylindole production from tryptophan by rumen microbes (Hammond and Carlson, 1980). Other work has indicated that monensin minimized lung lesions when cattle were dosed with tryptophan to induce pulmonary edema and emphysema (Boren et al., 1982).

Several studies have provided evidence that monensin influences the precursors used for glucose synthesis and glucose turnover. While these are not ruminal events, it appears that the production of propionate in the rumen is causing these responses. It has been demonstrated that added propionate increases gluconeogenesis and body glucose turnover (Judson and Leng, 1973). It has also been demonstrated that the feeding of monensin will increase body glucose turnover by 14% (Van Maanen et al., 1978). Studies concerning amino acids as gluconeogenic precursors have indicated that monensin decreases the amount of amino acid that is used for glucose synthesis (Beede et al., 1980). These studies support the concept that monensin functions to increase propionate production in the rumen which provides more of this substrate for gluconeogenesis and thereby decreases the use of amino acids for this purpose. This phenomemon relates to the previously discussed protein-sparing effect as well as indicates a potentially important difference in glucose metabolism.

Research has indicated that feeding monensin to heifers results in earlier puberty (McCartor et al., 1979). These studies have been conducted in several ways to demonstrate that the puberty response is independent of weight of the animal. While the mode of action causing this response is not clear, propionate infusion studies with prepuberal heifers (Rutter et al., 1981) have indicated an increased release of luteinizing hormone resulting from a gonadotropin releasing hormone challenge. Thus, it is possible that propionate may again be playing an important biological role.

A recently reported response due to feeding monensin is that of reduced face fly and horn fly numbers (Herald et al., 1982). This work indicated that monensin reduced face fly and horn fly pupae in cattle feces by 20% and 23% respectively. The surviving pupae were also smaller which resulted in a decreased potential egg production. There is no evidence as to the mode of action, but it is possible that intraruminal changes due to feeding monensin alter the composition of the feces in such a way that it is detrimental to larval growth of these insects.

REFERENCES

Anonymous. 1975a. Rumensin Research Seminar. Eli Lilly and Co., Indianapolis, IN.

Anonymous. 1975b. Rumensin Technical Manual. Elanco Products Co., Indianapolis, IN.

Anonymous. 1982. Avotan 50 Growth Promoter Product Manual. SA Cyanamid Ltd., Johannesburg, South Africa.

Baldwin, R.L., Koong, L.J. and Ulyatt, M.J. 1977. A dynamic model of ruminant digestion for evaluation of factors affecting nutritive value. Agric. Syst. 2:255-288.

Bartley, E.E., Herod, E.L., Bechtle, R.M., Sapienze, D.A., Brent, B.E. and Davidorich, D. 1979. Effect of monensin, lasalocid, or a new polyether antibiotic with and without niacin or amicloral on rumen fermentation in vitro and on heifer growth and feed efficiency. J. Anim. Sci. 49:1066-1075.

Beede, D.K. and Farlin, S.D. 1977. Effects of antibiotics on apparent studies. J. Anim. Sci. 45:385-392.

Beede, D.K., Gill, W.W., Koenig, S.E., Lindsey, T.O., Schelling, G.T., Mitchell, Jr., G.E. and Tucker, R.E. 1980a. Nitrogen utilization and fiber digestibility in growing steers fed a low protein diet with monensin. J. Anim. Sci. 51 (Suppl. 1): 5.

Beede, D.K., Trabue, P.J., Schelling, G.T., Michell, Jr., G.E. and Tucker, R.E. 1978. Nitrogen balance and energy digestibility in growing goats fed a low protein diet with or without momensin. J. Anim. Sci. 47 (Suppl. 1):114.

Beede, D.K., Schelling, G.T., Mitchell, Jr., G.E. and Tucker, R.E. 1980. Gluconeogenesis from threonine in growing goats abomasally administered glucose, propionate, oleate or fed monensin. J. Anim. Sci. 51 (Suppl. 1):345.

Boren, B.B., Schelling, G.T. and Ellis, W.C.. 1982. Tryptophan -induced pulmonary edema and emphysema in cattle fed a grain diet or alfalfa hay with and without monensin. J. Anim. Sci. 55 (Suppl. 1):51.

Bull, L.S., Rumpler, W.V., Sweeney, T.F. and Zinn, R.A. 1979. Influence of ruminal turnover on site and extent of digestion. Fed. Proc. 38:2713- 2719.

Chalupa, W. 1977. Manipulating rumen fermentation. J. Anim. Sci. 45:585-599.

Chalupa, W. 1980. Chemical control of rumen microbial metabolism. In: Y. Ruckebusch and P. Thivend (Eds.), Digestive Physiology and Metabolism in Ruminants. MTP Press Ltd., Lancaster England. p. 325-347.

Chalupa, W., Corbett, W. and Brethour, J.R. 1980. Effects of monensin and amicloral on rumen fermentation. J. Anim. Sci. 51:170-179.

Chen, M. and Wolin, M.J. 1978. Effect of monensin and lasalocid on the growth of rumen and methane bacteria. Abstr., Am. Soc. Microbial., p. 92.

Dartt, R.M., Boling, J.A. and Bradley, N.W. 1978. Supplemental protein withdrawal and monensin in corn silage diets for finishing steers. J. Anim. Sci. 46:345-349.

Dennis, S.M., Nagaraja, T.G. and Bartley, E.E. 1980. Effect of lasalocid or monensin on lactic acid production by rumen bacteria. J. Anim. Sci. 51 (Suppl. 1):96.

Dinius, D.A., Simpson, M.S. and Marsh, P.B. 1976. Effect of monensin fed with forage on digestion and the ruminal ecosystem of steers. J. Anim. Sci. 42:229-234.

Ellis, W.C. and Delaney, D.S. 1982. Intraruminal responses to monensin in cattle grazing forages. Beef Cattle Research in Texas, 1982, page 84, Texas Agric. Exp. Sta.

Fontenot, J.P., Webb, K.E. and Lucas, D.M. 1980. Effect of salinomycin on in vitro and in vivo ruminal volatile fatty acids. J. Anim. Sci. 51 (Suppl. 1):360.

Gill, D.R., Owens, F.N., Fent, R.W. and Fulton, R.K. 1979. Thiopeptin and roughage level for feedlot steers. J. Anim. Sci. 49:1145-1150.

Hammond, A.C. and Carlson, J.R. 1980. Inhibition of ruminal degradation of L-tryptophan to 3-methylindole in vitro. J. Anim. Sci. 51:207-214.

Hanson, T.L. and Klopfenstein, T.J. 1979. Monensin, protein source and protein levels for growing steers. J. Anim. Sci. 48:474-479.

Harrison, D.G., Beever, D.E., Thompson, D.J. and Osbourn, D.J. 1975. Manipulation of rumen fermentation in sheep by increasing the rate of flow of water from the rumen. J. Agric. Sci. 85:93-101.

Herald, F., Knapp, F.W., Brown, S. and Bradley, N.W. 1982. Efficacy of monensin as a cattle feed additive against the face fly and horn fly. J. Anim. Sci. 54:1128-1131.

Herod, E.L., Bartley, E.E., Davidovich, A., Bechtle, R.M., Sapienza, D.A. and Brent, B.E. 1979. Effect of adaptation to monensin or lasalocid on rumen fermentation in vitro and the effect of these drugs on heifer growth and feed efficiency. J. Anim. Sci. 49 (Suppl. 1):374.

Hungate, R.E. 1966. The rumen and its microbes. p. 245. Academic Press, Inc., N.Y., N.Y., p. 206-245.

Ingle, D.L., Darlymple, R.H. and Kiernan, J.A. 1978. Avoparcin, a ruminant growth promoting antibiotic. J. Anim. Sci. 47 (Suppl. 1):424.

Joyner, Jr., A.E., Brown, L.J., Fogg, T.J. and Rossi, R.T. 1979. Effect of monensin on growth, feed efficiency and energy metabolism of lambs. J. Anim. Sci. 48:1065-1069.

Judson, C.J. and Leng, R.A. 1973. Studies on the control of gluconeogenesis in sheep: Effect of propionate, casein and butyrate infusions. Br. J. Nutr. 29:175-182.

Lemenager, R.P., Owens, F.N., Shockey, B.J., Lusby, K.S. and Totusek, R. 1978. Monensin effects on rumen turnover rate, twenty-four hour VFA pattern, nitrogen components and cellulose disappearance. J. Anim. Sci. 47:255-261.

McCarthy, F.D., Bergen, W.G. and Hawkins, D.R. 1979. Protein sparing effect and performance of growing - finishing steers fed monensin. J. Anim. Sci. 49 (Suppl.):79.

McCartor, M.M., Randel, R.D. and Carroll, L.H. 1979. Dietary alteration of ruminal fermentation on efficiency of growth and onset of puberty in brangus heifers. J. Anim. Sci. 48:488-494.

McClure, W.H., Fontenot, J.P., Webb, K.E. and Lucas, D.M. 1980. Feedlot performance of cattle fed different salinomycin levels. J. Anim. Sci. 51 (Suppl. 1):380.

Muir, L.A., Rickles, E.L., Duquette, P.F. and Smith, G.E. 1981. Prevention of induced acidosis in cattle by thiopeptin. J. Anim. Sci. 52:635-643.

Nagaraja, T.G., Avery, T.B., Bartley, E.E., Roof, S.K. and Dayton, A.D. 1982. Effect of lasalocid, monensin or thiopeptin on lactic acidosis in cattle. J. Anim. Sci. 54:649-658.

Owens, F.N., Shockey, B.J., Fent, R.W. and Rust, S.R. 1978. Monensin and abomasal protein passage of steers. J. Anim. Sci. 47 (Suppl. 1):114.

Owens, F.N. 1980. Ionophore effect on utilization and metabolism of nutrients-ruminants. In 1980 Georgia Nutrition Conference. University of Georgia, Athens, Ga. p. 17-25.

Pond, K.P. and Ellis, W.C. 1981. Effects of Monensin on fecal output and voluntary intake of grazed coastal bermudagrass. Beef Cattle Research in Texas, 1981, Texas Agric. Exp. Sta. p. 31.

Poos, M.I., Hanson, T.L. and Klopfenstein, T.J. 1979. Monensin effects on diet digestibility, ruminal protein bypass and microbial protein synthesis. J. Anim. Sci. 48:1516-1524.

Potter, E.L., Cooley, C.O. and Richardson, L.F. 1979. Effect of narasin upon the performance of feedlot cattle. J. Anim. Sci. 49 (Suppl. 1):397.

Pressman, B.C. 1976. Biological applications of ionophores. Anim. Rev. Biochem. 45:501-530.

Raun, A.P., Cooley, C.O., Potter, E.L., Rathmacher, R.P. and Richardson, L.F. 1976. Effect of monensin on feed efficiency of feedlot cattle. J. Anim. Sci. 43:670-677.

Richardson, L.F., Raun, A.P., Potter, E.L. and Cooley, C.O. 1976. Effect of Monensin in rumen fermentation in vitro and in vivo. J. Anim. Sci. 43:657-664.

Richardson, L.F., Potter, E.L. and Cooley, C.O. 1978. Effect of monensin on ruminal protozoa and volatile fatty acids. J. Anim. Sci. 47 (Suppl. 1):45.

Rust, S.R., Owens, F.N., Thornton, J.H. and Fent, R.W. 1978. Monensin and digestibility of feedlot rations. J. Anim. Sci. 47 (Suppl. 1):437.

Rutter, L.M., Randel, R.D., Schelling, G.T. and Forrest, D.W. 1981. Effect of abomasal infusion of propionate on GNRH-induced luteinizing hormone release in prepuberal heifers. J. Anim. Sci. 53 (Suppl. 1):364.

Schanne, F.A.X., Kane, A.B., Young, E.E. and Farber, J.L. 1979. Calcium dependence of toxic cell death: a final common pathway. Science 206:700-701.

Schelling, G.T., Spires, H.R., Mitchell, Jr., G.E. and Tucker, R.E. 1977. The effect of various antimicrobials on amino acid degradation rates by rumen microbes. Fed. Proc. 37:411.

Short, D.E., Bryant, M.P., Hinds, F.C. and Fahey, G.C. 1978. Effect of monensin upon fermentation end products and cell yield of anaerobic microorganisms. J. Anim. Sci. 47 (Suppl. 1):44.

Simpson, M.E. 1978. Effects of certain antibiotics on in vitro cellulose digestibility and volatile fatty acid production by ruminal microorganisms. J. Anim. Sci. 47 (Suppl. 1):429.

Smith, G.E. 1971. Energy metabolism. In: D. Church (Ed.), Digestive physiology and nutrition of ruminants, first edition. O and B Books, inc., Corvalis, OR p. 601-615.

Smith, J.B. and Rosengurt, E. 1978. Serum stimulates the Na-K pump in quiescent fibroblasts by increasing Na entry. Proc. National Acad. Sci. 75:5560-5565.

Stuart, R.L. and Zimmerman, C.R. (Eds.). 1982. Bovatec Symposium Proceedings. Hoffman-La Roche Inc., Nutley, New Jersey.

Thomson, D.J., Beever, D.E., Latham, M.J., Sharpe, M.E. and Terry, R.A. 1978. The effect of inclusion of mineral salts in the diet on dilution rate, pattern of rumen fermentation and the composition of the rumen microflora. J. Agric. Sci. 91:1-7.

Thornton, J.H. and Owens, F.N. 1981. Monensin supplementation and in vivo methane production by steers. J. Anim. Sci. 52:628-634.

Trei, J.E., Singh, J.K. and Scott, G.C. 1970. Effect of methane inhibitors on rumen metabolism. J. Anim. Sci. 31:256.

Van Maanan, R.W., Herbein, J.H., McGilliard, A.D. and Young, J.W. 1978. Effects of monensin in in vivo rumen propionate production and blood glucose kinetics in cattle. J. Nutr. 108:1002-1007..

Van Nevel, C.J. and Demeyer, D.I. 1977. Effect of monensin on rumen metabolism in vitro. Appl. Environ. Microbial. 34:251-257.

DISCUSSION ON DR. SCHELLING'S PAPER

Dr. Meyer: Firstly does an increase in propionate affect carcass composition? Secondly, since ionophores have toxic effects, is the level fed and the toxic level close?

Dr. Schelling: Evidence shows no change in fat to lean ratio in the carcass of treated animals. The toxic level of ionophores varies among species. There is no problem with monensin in cattle since it is not very toxic and high levels in the ration depress feed intake and this is a built in safety mechanism. However this does not apply to the horse. Monensin is very toxic to the horse. In general, with lasalocid there appears to be a wider safety range.

Dr. Heitzman: Nevertheless several cases of deaths in cattle due to excessive doses of monensin have been reported.

Dr. Beinfait: Do we know the metabolites and metabolism of monensin?

Dr. Schelling: There is a significant absorption of monensin. Work by Eli Lilly and Co. suggests that at least one third of it is absorbed into the body proper and re-excreted by the bile back into the gastrointestinal tract. That proportion that is absorbed is metabolised to a considerable extent - I am not familiar with all the metabolites but significant amounts are absorbed, metabolised and efficiently re-excreted back into the G I tract.

Dr. Roche: What is the mechanism by which monensin affects the release of LH?

Dr. Schelling: I do not know. Work indicated that feeding monensin increased LH episodic peaks. This was followed up with a propionate infusion study with prepubertal heifers and much the same response was seen. The infusion experiment effect may be an energy effect since we added propionate and therefore more energy, or it could be something specific to propionate.

Dr. Raymond: You said that pasture intake was measured indirectly, and you used an estimate of digestibility, but monensin may affect digestibility. I wonder how much of the effect on pasture intake which you reported might be an artifact because proper adjustments for the effect of monensin on digestibility were not made.

Dr. Schelling: Adjustments on this were made by conducting in vitro studies on forage, so that factor has theoretically been taken out. But I agree it is subject to large error. Looking at the data I would not necessarily view it as a 15% response, but it is indicitive of a positive response, whatever the magnitude.

Dr. Verbeke: Are there additive effects with anabolic agents and monensin considering they have different modes of action?

Dr. Schelling: Anabolic agents and monensin are used routinely in our feedlots in the U.S. and approximate additive effects are obtained.

Dr. Heitzman: I think Dr. Roche has also reported similar results.

Dr. Galbraith: You have shown a very large number of physiological responses to monesin and other ionophores. Can you relate these responses to the known biochemical effects of monensin and other ionophores at the microbial cell membrane?

Dr. Schelling: People have not really looked at the rumen microbes in the same basic sense with regard to transport of ions across membranes as they have with mouse fibroblasts. All the fundamental work with ionophores that I am aware of has been done with mammalian type cells. We are assuming that these basic ionic effects and transport systems pertain to microbial cells as well.

Dr. Galbraith: I was under the impression that ionophores are active against gram positive bacteria but you showed that streptocci and lactobacilli were not affected by monensin at the concentrations used.

Dr. Schelling: In general your comment on gram positives is correct but there does appear to be exceptions.

Dr. Tsakalof: Would you have any information on the role of monensin in preventing or curing acetonemia?   Do you have any information on Vitamin D complex reducing flora in the rumen?

Dr. Schelling: No to both questions.

THE MODE OF ACTION OF ANABOLIC AGENTS

with special reference to their effects

on protein metabolism - some speculations.

Peter J. Buttery and Patrick A. Sinnett-Smith

Department of Applied Biochemistry & Food Science,
University of Nottingham School of Agriculture,
Sutton Bonington, Nr. Loughborough, Leics. LE12 5RD.

ABSTRACT

Anabolic preparations which act at the tissue level can often be classified into those with androgenic properties and those with oestrogenic properties, however, it may not be these properties which are responsible for their growth promoting ability. The mode of action of the anabolic agent trenbolone, with androgenic properties, and the anabolic agent zeranol, with oestrogenic properties, is discussed. It is considered unlikely that trenbolone is acting as a classical androgen, at least as far as its action on muscle is concerned. Zeranol is considered unlikely to be acting solely via growth hormone. The question is posed: is it possible that these two agents just cause slight changes in the total endogenous hormone pattern which is sufficient to account for their growth promoting properties.

INTRODUCTION

Growth stimulation following treatment with anabolic agents is normally associated with an increase in the rate of protein deposition. The mode of action of these agents upon protein metabolism is far from being clear and at present there is little concrete evidence on which to construct a detailed hypothesis. These agents presumably act by either altering the endogenous hormone activities of the animal, by a direct action following binding to target tissue receptors, by antagonising the binding of endogenous hormones to their receptors, or by a combination of these processes. This review will concentrate on two agents - trenbolone (with androgenic properties) and zeranol (with relatively weak oestrogenic properties).

HORMONAL CONTROL OF PROTEIN DEPOSITION

Protein metabolism in animals is subject to considerable hormonal control, particularly in muscle. Variation
in amino acid supply over physiological ranges has little
effect upon muscle protein synthetic rates (Li and
Jefferson, 1978). There are numerous hormones which have
marked effects on protein metabolism and their action is
outlined in Table 1. Besides these direct effects, many
of these hormones interact with each other. For example,
in diabetic rats the effects of exogenously applied insulin
on muscle protein synthesis are masked by the elevated
corticosterone concentration which is found in these
animals. Oestrogens have been said to exert their influence
on growth by the secondary action of other hormones, e.g.
an increase in growth hormone or a change in the thyroid
status of the animal (see Heitzman, 1981). Growth hormone
messenger RNA synthesis is influenced by T_3 and
glucocorticoids (Shapi et al, 1978).

Table 1.   Effect of Hormones on Protein Metabolism,
especially Muscle.

Hormone	Effect
Androgens	Generally anabolic, promote protein synthesis.
Corticosteroids	Catabolic, reducing synthesis and degradation but main effect on synthesis. Tends to increase proportion of fat in carcass.
Oestrogens	Depends on species and dose. Tend to be anabolic in ruminants. In rats tend to be catabolic but small doses may be anabolic.
Thyroxine Thyroid hormones	Tend to be catabolic, stimulate synthesis but degradation stimu- lated more. Exact response very dependent upon dose.
Glucagon	Promotes production of carbohydrate from amino acids thereby diverting amino acids from protein synthesis.
Insulin	Promote synthesis, overall effect anabolic.

213

Hormone	Effect
Growth Hormone/ somatomedin	Promote synthesis, overall effect anabolic.
Peptide growth factors	Promote syntheses.

For more details see Buttery, 1983.

EFFECTS OF ANDROGENIC AGENTS ON PROTEIN METABOLISM

Testosterone is said to stimulate muscle protein synthesis although there are few in vivo experiments which have clearly shown this (see Wade and Gray, 1979). This action would appear to be due to a direct action on the muscle. Since there are specific androgen receptors in muscle (rat and pig Snochowski et al, 1980, 1981; sheep, Sinnett-Smith and Buttery, unpublished). In skeletal muscle and kidney there is no 5α reductase activity (Bardin et al, 1978). In some reproductive tissues (e.g. prostate) testosterone is a prehormone which must first be reduced to dihydrotestosterone which is then bound by the androgen receptor before stimulating the tissue (Mainwaring, 1977). It is also possible that testosterone might exert some of its action following aromatization to oestradiol (Gentry and Wade, 1976). Testosterone treatment reduces body fat and adipose tissue contains the enzymes to aromatize testosterone.

Trenbolone has androgenic properties but when given to female rats, mice and sheep reduces the rate of muscle protein synthesis rats - gastrocnemius muscle - (Vernon and Buttery, 1976, 1978a, 1978b); mice - (Carrington and Ward, 1981); and sheep - L-dorsi (Sinnett-Smith et al, 1983). To promote an increase in muscle mass this reduction in synthesis must be accompanied by a decrease in the rate of protein breakdown. N^tmethylhistidine excretion is decreased and muscle cathepsin D activity reduced in the rat (Vernon and Buttery, 1978b, 1981). Studies measuring the loss of label from muscle protein have also confirmed this (Vernon and Buttery (1976). Cathepsin D activity in sheep muscle is also decreased

(Sinnett-Smith et al, 1983 see Table 2). These changes are not consistent with the effects of trenbolone on muscle metabolism being directly due to its androgenic properties.

Although many of the studies on the mode of action of trenbolone have used the female laboratory rat, it is essential that such work is confirmed using a target species since there are differences between the metabolism of trenbolone between the rat and the heifer (Pottier et al, 1981).

With trenbolone, effects on growth rate of entire males or castrate males are small unless some oestrogenic compound is present. Females respond well to trenbolone alone. Any hypothesis which attempts to describe the mode of action of this agent must account for this sexual dimorphism.

Trenbolone treatment has been shown to cause a wide variety of changes in circulating hormone concentrations and some of these are summarized in Table 3.

The thyroid hormones have a marked effect on protein metabolism in muscle. The response obtained is very dependent upon the dose rate and the muscle studied. Hyperthyroidism is associated with a decreased skeletal mass. Muscle protein degradation falls during thyroidectomy and is restored on T_3 treatment, at least in rats (Brown et al, 1981). While this restoration of protein degradation rate by T_3 treatments is accompanied by an increase in muscle protein synthesis, a further increase in the availability of thyroid hormones results in even more protein catabolism uncompensated by increased protein synthesis (Goldberg et al, 1980). Thyroid hormone status has been implicated in the control of growth for many years, indeed some of the early anabolic agents were thyrostatic agents (Burroughs et al, 1958). Treatment of wethers, steers, and heifers with TBA generally reduces circulating thyroid hormone concentration (Table 3) and this would be

TABLE 2.   The Effects of TBA or Zeranol Treatment
            on Entire Female Sheep.

	Control	TBA	Zeranol	Pooled S.E.D.	Significance of comparison	
					CVT	CVZ
Weight Gain (kg)	5.8(5)	10.7(6)	9.4(6)	1.49	P<0.01	P<0.05
Muscle Protein FSR (d^{-1}) - H	0.060(5)	0.039(6)	0.041(5)	0.010	P<0.10	NS
Muscle Protein FSR (d^{-1}) - P	0.049(5)	0.033(5)	0.028(5)	0.0073	P<0.10	P<0.02
Muscle Cathepsin D Activity (Free)	3.79(5)	2.06(4)	2.00(3)	0.70	P<0.05	P<0.05
(Total)	15.05(5)	14.02(5)	16.00(6)	1.74	NS	NS

Number of animals per treatment in parenthesis.
Animals were slaughtered 4 weeks after treatment with TBA
(80mg) or Zeranol (12mg). Fractional synthetic rate (FSR)
was measured using H^3 Leucine by the technique of Garlick,
et al, 1973 and calculated using either homogenate (H)
or plasma (P) specific activities (from Sinnett-Smith
et al, 1983).

TABLE 3   Hormonal Changes after Treatment with TBA alone and in Combination with other Agents.

Treatment	Animal	Growth Hormone	Pro-lactin	Insu-lin	Other Hormones	References
TBA	Wether	-		NC	-	
TBA + Oe	Lamb	NC	NC		T -	Donaldson et al, 1981
Oe		+		+	NC	
TBA + Hex	Bull	NC	NC	NC		Galbraith, 1979
TBA	Cull Beef Cow	NC	NC			Galbraith, 1980a
TBA	Beef Heifer	NC	NC	NC		Galbraith, 1980b
TBA	Beef Heifer	NC	NC			Galbraith & Miller, 1977
TBA		NC		NC		
TBA + Hex	Steers	+		NC		Galbraith & Watson 1978
Hex		+		NC		
TBA	Female Sheep				Oe +	McVinish et al, 1982
TBA + Hex	Steer	NC		NC		Galbraith & Coelho, 1978
	Bull	NC		NC		
TBA + Hex	Steer	NC	NC	NC		Galbraith & Geraghty, 1978
TBA	Steer	NC	NC	NC	T -	Heitzman et al, 1977
TBA + Oe		NC	NC	NC	-	
TBA + Hex	Steer	+	NC	NC		Peters & Reed, 1982
TBA + Oe	Steer				T -	Heitzman, et al, 1980
TBA					-	
TBA	Heifer				T -	Heitzman, et al, 1980
TBA	Female lambs				C -	Thomas & Rodway, 1982
TBA	Female lambs				Oe +	Sinnett-Smith & Buttery (unpublished observations)

NC   no significant change   +   increased plasma concentration

–   decreased plasma concentration

Oe,   Oestradiol   Hex, Hexoestrol

T,   Thyroxine   C,   Cortisol   TBA, Trenbolone acetate

consistent with the reduction in protein synthetic rate
seen in sheep. In rats however, we have not been able to
confirm these observations (see Table 4) but in these
animals muscle protein turnover is also reduced.

TABLE 4.   Plasma T_3 and T_4 Concentrations in the
Female Rat after Treatment with TBA.

	Control		TBA	
Initial wt (g)	121.6 ± 0.9	(10)	121.7 ± 0.2	(10)
Final wt (g)	147.3 ± 1.9	(10)	165.0 ± 2.5	(10)
Wt. Gain (g)	25.7 ± 1.3	(10)	44.2 ± 2.4	(10) ***
T_3 (ng/ml)	2.24 ± 0.18	(9)	2.27 ± 0.22	(9)
T_4 (µg/100ml)	3.39 ± 1.15	(9)	4.24 ± 1.51	(9)

Number of animals per treatment in parenthesis
*** Significantly different at P<0.001
Animals were treated with TBA (800µg/kg body weight) or an
oil placebo daily for 14 days.

These observations must therefore cast doubt on the
universal involvement of thyroid hormones in the action
of trenbolone, although it must be admitted that most of
these studies have been confined to total thyroid hormone
concentrations in the plasma and have paid little attention
to the free thyroid hormone concentrations. The involvement of the thyroid gland in growth promotion following
trenbolone still requires further investigation.

An attractive hypothesis to account for the reduction
in protein synthesis in muscle tissue on giving trenbolone
is that the agent interferes with the catabolic action of
glucocorticoids. To accept this hypothesis it is necessary
to accept that growth is normally being suppressed by the
presence of glucocorticoids. This is a difficult point
to prove: however, there is evidence from clinical studies
with children that do give some support to it (Van den
Brande et al, 1979). There are few studies on farm animals
which can be used as evidence. There is some suggestion

218

that growth rate is universally related to circulating
cortisol concentrations. Purchas et al, (1980) obtained
a negative correlation between circulating corticosteroid
concentration and growth rate of sheep and cattle. There
are other reports with cattle (e.g. Trenkle and Topel, 1978)
which confirm this but there are also instances where no
correlation has been obtained (e.g. Lange and Lindermann,
1972). What is clear is that glucocorticoid injection,
at least in cattle, increases fat content while reducing
protein content of the carcass (Carroll et al, 1963).
There is a tendency for trenbolone to have the opposite
effect (Sinnett-Smith et al, 1983). Mayer and Rosen (1975)
suggested that androgens may compete for the glucocorticoid
receptor of muscle and thus reduce the catabolic effects
of glucocorticoids. In more recent studies, however, some
considerable doubt has been cast on this suggestion by
Snochowski et al, (1980, 1981) who showed that in both
rat and porcine muscle there were distinct glucocorticoid
and androgenic receptors and there was little evidence of
any cross binding. Trenbolone does not bind to the glucocorticoid receptor of thymus glands taken from
 adrenalectomized rats (Raynaud et al, 1981 - see Table 5).
It therefore appears unlikely that binding to the
glucocorticoid receptor by trenbolone could account for
its effects on muscle protein synthesis and degradation
as was originally proposed by Buttery et al, (1978).

     An alternative influence of trenbolone on the action
of glucocorticoids has been suggested by Thomas and Rodway
(1982) who have shown that circulating cortisol
(corticosterone) concentrations are reduced by trenbolone
treatment of sheep and rats, especially at the diurnal
peaks (cortisol and corticosterone concentrations show a
marked diurnal rhythm). This effect has been traced to
an inhibition of glucocorticoid production by the adrenals.
This is, however, not specific for trenbolone (Thomas and
Rodway, 1982) and has been reported previously for other

steroids (Matsumara, 1976). Dexamethasone has also been shown to inhibit steroidogenesis following ACTH treatment of the adrenals (Guillemant and Guillemant, 1982). There are, however, other problems with this hypothesis since it would appear that the main effect of glucocorticoids on muscle protein turnover (for review see Buttery, 1983) is on the synthetic rate, while trenbolone appears to affect degradative rate more than synthetic rate.

Despite these objections there is other evidence which gives further support for the involvement of glucocorticoid status in trenbolone action. Trenbolone reduces female rat liver tyrosine transaminase activity (Rodway and Galbraith, 1979), an enzyme induced by glucocorticoids. In recent studies in these laboratories we have confirmed this observation and shown that with. trenbolone there is little change in activity in male rats treated with trenbolone (see Table 6). It is interesting to note that treated male rats show little, or in this case no, increase in growth rate. It is clear that the possible involvement of glucocorticoids in the mode of action of trenbolone is worthy of further investigation.

With trenbolone in combination with oestrogen there is some evidence for an elevated plasma growth hormone concentration. In heifers treated with trenbolone alone (Table 3) there is no evidence of any change. Although we have not tested this with the female rat we have measured circulating somatomedin concentrations using a pig costal cartilage assay (Coates et al, 1977) but have found no changes (Vernon, B.G. and Buttery, P.J., unpublished observations).

An interesting observation is that trenbolone apparently reduces the rate of metabolic clearance of ^{3}H when ^{3}H-oestradiol is implanted with trenbolone in a conventional manner (i.e. under the skin behind the ear). Another observation which may bear on this is that trenbolone apparently increases endogenous oestradiol concentrations

TABLE 5.   Binding of trenbolone to various receptors

Receptor	Conditions of incubation	Source of Receptor	Binding of relative to reference hormone (100)	Reference hormone
Oestrogen	2hr.0°	immature mouse uteri	.1	oestradiol
Progestin	2hr.0°	oestradiol primed rabbit uteri	75	progesterone
	24hr.0°		15	
Androgen	30min.0°	Castrated rat prostates	250	testosterone
	2hr.0°		190	
Gluco-corticoid	1hr.0°	Adrena-lectomised rat thymuses	9	corticosterone
	24hr.0°		2.5	

Taken from Raynaud _et al_, (1981)

TABLE 6   The Effect of TBA on Rat Liver TAT Activity where animals were treated with TBA (800 µg/ kg body wt/day or oil for 16 days.

	Initial wt. (g)	Final wt. (g)	TAT Activity (µmol/min/g liver)
Males			
Control	122.2 ± 0.89	207.8 ± 1.83	6.18 ± 0.52
TBA	122.5 ± 1.14	194.9 ± 2.89**	6.57 ± 0.51
Females			
Control	108.6 ± 2.14	148.35 ± 2.86	8.28 ± 0.63
TBA	108.5 ± 1.71	170.8 ± 3.69***	6.39 ± 0.54*

P.A. Sinnett-Smith, A. Spencer and P. J. Buttery (unpublished observations)

Significance of difference from controls *$P < 0.05$, **$P < 0.01$, ***$P < 0.001$

TAT = Tyrosine transaminase

in female lambs given trenbolone (MacVinish et al, 1983;
Sinnett-Smith and Buttery, unpublished observations).
Perhaps it should be stated that what was being measured
during these studies was binding to an oestradiol antibody. Interestingly, there are some metabolites of
trenbolone which have aromatized "A" rings (Pottier et al,
1981). Is it possible that the effects of trenbolone
on muscle may be due to some oestrogenic compound,
either from trenbolone itself or the concentration of
which is being increased by the action of trenbolone?
Before leaving consideration of the mode of action of
trenbolone it should be remembered that trenbolone does
bind to a variety of receptors (see Table 5) and could be
working via an agonistic or antagonistic action on one
of these receptors.

MODE OF ACTION OF AN AGENT WITH OESTROGENIC
PROPERTIES, ZERANOL

Zeranol (Ralgro) has been employed for growth
promotion for a longer time than trenbolone but there
have been few studies on its mode of action. Again
this substance causes a variety of changes in circulating
hormone concentrations. Of particular note is the
elevation in growth hormone, see Table 7. This is a
common observation associated with the treatment of
animals, especially ruminants, with oestrogens and it is
often said that oestrogens cause their growth promotional
effects by elevating growth hormone. In studies in a
laboratory we have investigated the effects of zeranol
on the protein synthesis of muscle in female lambs. Our
data (Table 2) give no indication that protein synthesis
is elevated, if anything there is a reduction. If
growth hormone were responsible for the action of zeranol,
then protein synthesis would have been expected to be
elevated. Female rats have a slower rate of protein
synthesis than males, although it should be remembered
that they grow slower (Waterlow and Stephen, 1968). Rat

muscle does contain receptors for oestrogens (Dahlberg, 1982)
and we have recently confirmed their presence in sheep
muscle (Palmer et al, unpublished observations). Zeranol
has also been shown to inhibit oestradiol binding to its
uterine receptor (Peck and Chesworth, 1977) and to its
hepatic receptor (Powell-Jones et al, 1981). The
possibility of a direct action on muscle receptors cannot
be totally excluded. There is some evidence that zeranol
reduces circulating T_3 concentrations (see Table 7). Thus
again it is possible that thyroid status is important in
the mode of action of this agent.

The question therefore has to be asked, does zeranol
or a metabolite act directly upon muscle? Another question;
is it possible that zeranol (and oestrogenic compounds in
general) have a mode of action which is similar to that
of trenbolone? Heitzman et al, (1981) studied the interaction between oestrogens (hexoestrol or oestradiol) and
trenbolone on the growth rate of steers. Their conclusion
was that the effects were independent and additive.
Heitzman (1981) did, however, point out that there was
evidence which points to similarities in the mode of action
of androgens and oestrogens. Trenbolone plus zeranol is
a good growth promoting combination for castrate male
cattle.

CONCLUSION

At present it is difficult to provide a convincing
hypothesis which accounts for the mode of action of the two
agents, zeranol and trenbolone. One answer might be
that they do not act via a single mechanism but that both
agents act by making small changes to endogenous hormone
concentrations, the net result of which is increased growth.
Such a mechanism may also be accompanied by a direct action
of the agents on the target tissues. The possibility has
also to be considered that the primary action is not on
muscle protein metabolism but is on another tissue and that
the effects on muscle are of a secondary nature.

TABLE 7   Hormonal Changes after treatment with Zeranol

Animal	Growth Hormone	Insulin	Other Hormones	References
Steer	+			Borger et al, 1973a
Steer	+	NC		Borger et al, 1971
Steer	+	NC		Borger et al, 1973b
Steer	+			Fuller et al, 1981
Heifer	+		LH NC	
Heifer	+	+		Galbraith, 1979
Steer			T_4 - T_3 -	Gopinath & Kitta, 1981
Steer (iv)	NC	NC	C NC	Heath et al, 1980
Wether lamb (iv)	NC	NC		Olsen et al, 1977
Wether lamb	+	+		
Ram lambs			C -	Riesen et al, 1977
Wether lambs			FSH -	
Steer		+		Sharp & Dyer, 1970
Wether lambs		+		Wangness et al, 1981
Wether lambs			T_4 -	Wiggins et al, 1980
Female lambs			Oe NC	Sinnett Smith & Buttery (unpublished observations)

NC   no significant change
+   increased plasma concentration
-    decreased plasma concentration
LH   Luteinizing hormone
(iv) intravenous infusion.   Zeranol was implanted unless otherwise stated.

FSH   Follicle stimulating hormone
Oe   Oestradiol
C   Cortisol

The Agricultural Research Council is thanked for support for the authors studies reported in this review. Christine Palmer and Christine Essex are thanked for expert technical assistance.

REFERENCES

Bardin, C.W., Bullock, L.P., Mills, N.C., Lin, Y-C and Jacob, S.T. 1978. The role of receptors in the anabolic action of androgens. In "Receptors and Hormone Action II" (Ed. B.W. O'Malley and L. Birnbaumer) Academic Press, New York. pp.83-103.

Borger, M.L., Wilson, L.L., Sink, J.D., Ziegler, J.H., Davis, S.L., Orley, C.F. and Rugh, M.C. 1971. Zeranol and protein effects on finishing steers. J Anim. Sci., 33, 275-276.

Borger, M.L., Sink, J.D., Wilson, L.L., Ziegler, J.H. and Davis, S.L. 1973a. Zeranol and dietary protein level effects on DNA, RNA and protein composition of 3 muscles and the relationships to serum insulin and growth hormone levels of steers. J. Anim. Sci. 36, 712.

Borger, M.L., Wilson, L.L., Sink, J.D., Ziegler, J.H. and Davis, S.L. 1973b. Zeranol and dietary protein level effects on live performance, carcass merit, certain endocrine factors and blood metabolite levels of steers. J. Anim. Sc., 36, 706-711.

Brown, J.G., Bates, P.C. Holliday, M.A. and Millward, D.J. 1981. Thyroid hormones and muscle protein turnover. Biochem. J. 194, 771-782.

Burroughs, W., Cheng, E., Raun, A., Culbertson, C.C. and Kline, E.A. 1958. Beneficial influences of methimazole in fattening cattle rations. J. Anim. Sci., 17, 1163.

Buttery, P.J. 1983. Hormonal control of protein deposition. Proc. Nutr. Soc., In Press.

Buttery, P.J., Vernon, B.G. and Pearson, J.T. 1978. Anabolic agents - some thoughts on their mode of action. Proc. Nutr. Soc., 37, 311-315.

Carrington, L.E. and Ward, L.C. 1981. Tissue protein synthetic activity in trenbolone acetate treated mice. Nutr. Rep. Intern. 24, 1199-1204.

Carroll, F.D., Powers, S.B. and Clegg, M.T. 1963. Effect of cortisone acetate on steers. J. Anim. Sci., 22, 1009-1011.

Coates, C.L., Burwell, R.G., Buttery, P.J. and Woodward, P. 1977. Somatomedin activity in synovial fluid. Annals Rheumatic Diseases, 36, 50-55.

Dahlberg, E. 1982. Characterization of the cytosolic estrogen receptor in rat skeletal muscle. Biochim. Biophys. Acta., 717, 65-75.

Donaldson, I.A., Hart, I.C. and Heitzman, R.J. 1981. Growth hormone, insulin, prolactin and total thyroxine in plasma of sheep implanted with the anabolic steroid trenbolone acetate alone or with oestradiol. Res. Vet. J., 30, 7-13.

Fuller, T.S., Kaltenbach, C.C. and Dunn, T.G. 1981. Endocrine patterns following zeranol implantation in beef cattle. J. Anim. Sci., 53, Suppl., 319.

Galbraith, H. 1979. Growth, metabolic and hormonal response in the blood of British Fresian entire male cattle treated with trenbolone acetate and hexoestrol. Anim. Prod. 28, 417.

Galbraith, H. 1980a. Effect of trenbolone acetate on growth, blood metabolites and hormones of cull beef cows. Vet. Rec., 107, 559-560.

Galbraith, H. 1980b. The effect of trenbolone acetate on growth, blood hormones and metabolites and on nitrogen balance in beef heifers. Anim. Prod., 30, 389-394.

Galbraith, H. and Coelho, J.F.S. 1978. The effect of dietary protein intake and implantation with trenbolone acetate and hexoestrol on growth performance and blood metabolites and hormones of British Fresian Steers. Anim. Prod., 26, 360.

Galbraith, H. and Geraghty, K.J. 1978. Effect of dietary energy intake and implantation with trenbolone acetate and hexoestrol on growth performance, blood metabolites and hormones of British Fresian steers. Anim. Prod., 26, 361.

Galbraith, H. and Miller, T.B. 1977. The effect of trenbolone acetate on the performance, blood metabolites and hormones and nitrogen metabolism of beef heifers. Anim. Prod., 24, 133.

Galbraith, H. and Watson, H.B. 1978. Performance, blood and carcass characteristics of finishing steers treated trenbolone acetate and hexoestrol. Vet. Rec., 103, 28-31.

Gentry, R.T. and Wade, G.N. 1976. Androgenic control of food intake and body weight in male rats. J. Comp. Physiol. Psychol., 90, 18-25.

Goldberg, A.L., Tischler, M., DeMartino, G. and Griffin, G. 1980. Hormonal regulation of protein degradation and synthesis in skeletal muscle. Fed. Proc., 39, 31-36.

Gopinath, R. and Kitts, W.D. 1981. Effect of anabolic compounds on plasma levels of thyroid hormones in growing beef steers. J. Anim. Sci., 53, Suppl., 321.

Guillemant, J. and Guillemant, S. 1982. Effect of dexa-methasone on steroidogenesis and on adrenocortical cyclic AMP and cyclic GMP responses to ACTH. Horm. metab. Res., 14, 547-550.

Heath, S.P., Herbein, J.H., Russell, R.W., McGilliard, A.D. and Young, J.W. 1980. Effects of intravenously infused zeranol on glucose kinetics in growing Holstein steers. J. Dairy Sci., 63, 553.

226

Heitzman, R.J. 1981. Mode of action of anabolic agents in hormones and metabolism in ruminants (Ed. J.M. Forbes and M.A. Lomax). Agricultural Research Council. pp. 129-139.

Heitzman, R.J., Chan, K.H. and Hart, I.C. 1977. Liveweight gains, blood levels of metabolites, proteins and hormones following implantation of ababolic agents in steers. Br. Vet. J., 133, 62-70.

Heitzman, R.J., Donaldson, I.A. and Hart, I.C. 1980. Effect of anabolic steroids on plasma thyroid hormones in steers and heifers. Br. Vet. J., 136, 168-174.

Heitzman, R.J., Gibbons, F.N., Little, W. and Harrison, L.P. 1981. A note on the comparative performance of beef steers implanted with anabolic steroids trenbolone acetate and oestradiol 17β alone or in combination. Anim. Prod., 32, 219-222.

Lange, W. and Lindermann, E. 1972. Investigation of the function of the adrenal cortex in fattening cattle of different genotypes. Arch. Tierz., 15, 171-188.

Li, J.B. and Jefferson, L.S. 1978. Influence of amino acid availability on protein turnover in perfused sketetal muscle. Biochim. Biophys. Acta., 544, 351-359.

MacVinish, L.J., Galbraith, H. and Chesworth, J.M., 1983. Steroid concentrations in sheep implanted with trenbolone actate. Proc. Nutr. Soc. In Press.

Mainwaring, W.I.P. 1977. The mechanism of action of androgens. Springer-Verlag, New York.

Matsumara, M. 1976. Pharmacological studies on adrenal cortex suppression caused by anabolic steroid. J. Kansai Med. Univ., 28, 32-44.

Mayer, M. and Rosen, F. 1975. Interactions of anabolic steroids with glucocorticoid receptor sites in rat muscle cytosol. Am. J. Physiol., 229, 1381-1386.

Olsen, R.F., Wangness, P.J., Martin, R.J. and Gahagan, J.H. 1977. Effects of zeranol on blood metabolites and hormones in wether lambs. J. Anim. Sci., 45, 1392.

Peck, D.N. and Chesworth, J.M. 1977. Estrogenic activity of zeranol in ewes. Horm. Metab. Res., 9, 531-532.

Peters, A.R. and Read, d.J. 1982. Effect of trenbolone actate and hexoestrol on plasma hormone and metabolite concentrations in steers. Anim. Prod., 34, 395.

Powell-Jones, W., Raeford, S. and Lucier, G.W. 1981. Binding properties of Zeranalenone mycotoxin to hepatic estrogen receptors. Molec. Pharmacol., 20, 35-42.

Pottier, J., Cousty, C., Heitzman, R.J. and Reynolds, I.P. 1981. Differences in the biotransformation of a 17β-hydroxylated steroid, trenbolone acetate, in rat and cow. Xenobiotica, 11, 489-500.

Purchas, R.W., Barton, R.A. and Kirton, A.H. 1980. Relationships of circulating cortisol levels with growth rate and meat tenderness of cattle and sheep. Aust. J. Agric. Res., 31, 221-232.

Raynaud, J.P., Ojasoo, T. and Labrie, F. 1981. Steroid hormones - agonists and antagonists. In "Mechanisms of Steroid Action" (Ed. G.P. Lewis and M. Ginsburg). MacMillan, London, pp. 145-157.

Riesen, J. W., Beeler, B.J., Abenes, F.B. and Woody, C.O. 1977. Effect of zeranol on reproductive system of lambs. J. Anim. Sci., 45, 293.

Rodway, R.G. and Galbraith, H. 1979. Effects of anabolic steroids on hepatic enzymes of amino acid catabolism. Horm. Metab. Res., 11, 489-490.

Shapiro, L.E., Samuels, H.H. and Yaffe, B.M. 1978. Thyroid and glucocorticoid hormones synergistically control growth hormone mRNA in cultured GH_1 cells. Proc. Natn. Acad. Sci., U.S.A., 75, 45-49.

Sharp, G.D. and Dyer, I.A. 1970. Metabolic responses to zearalanol implants. J. Anim. Sci., 30, 1040.

Sinnett- Smith, P.A., Dumelow, N.W. and Buttery, P.J. 1983. The effects of trenbolone acetate and zeranol on protein metabolism in male castrate and female lambs. Br. J. Nutr. In Press.

Snochowski, M., Dahlberg, E. and Gustafsson, J.A. 1980. Characterization and quantification of the androgen and glucocorticoid receptors in cytosol from rat skeletal muscle. Eur. J. Biochem., 111, 603-616.

Snochowski, M., Lundstrom, K., Dahlberg, E., Petersson, H., and Edquist, L.E. 1981. Androgen and glucocorticoid receptors in porcine skeletal muscle. J. Anim. Sci., 53, 80-90.

Thomas, K.M. and Rodway, R.G. 1982. Effects of trenbolone on corticosterone production by isolated rat adrenal cells. Proc. Nutr. Soc., In Press.

Trenkle, A. and Topel, D.G. 1978. Relationships of some endocrine measurements to growth and carcass composition of cattle. J. Anim. Sci., 46, 1604-1609.

Van den Brande, J.L., Van Buul-Offers, S.C., DuCaju, M.V.L., Price, D.A., Wilt, J.M. and Bongers-Schokking, J.J. 1979. Somatomedin and the regulation of statural growth. In "Somatomedins and Growth" (Ed. G. Giordano, J.J. Van Wyk and F. Minuto). Academic Press, London pp. 255-262.

Vernon, B.G. and Buttery, P.J. 1976. Protein turnover in rats treated with trienbolone acetate. Br. J. Nutr., 36, 575-579.

Vernon, B.G. and Buttery, P.J. 1978a. The effect of trenbolone acetate with time on the various responses of protein synthesis in the rat. Br. J. Nutr., 40, 563-572.

Vernon, B.G. and Buttery, P.J. 1978b. Protein metabolism of of rats treated with trienbolone acetate. Anim. Prod., 26, 1-9.

Vernon, B.G. and Buttery, P.J. 1981. The effect of the growth promoter trenbolone acetate, dexamethasone and thyroxine on skeletal muscle Cathepsin D (EC3.4.4.23) activity. Proc. Nutr. Soc., 40, 13A.

Wade, G.N. and Gray, J.M. 1979. Gonadal effects on food intake and adiposity: a metabolic hypothesis. Physiol. Behav., 22, 583-593.

Wangness, P.J., Olsen, R.F. and Martin, R.J. 1981. Effects of breed and zeranol implantation on serum insulin, somatomedin-like activity and fibroblast proliferative activity. J. Anim. Sci., 52, 57-62.

Waterlow, J.C. and Stephen, J.M.L. 1968. The effect of low protein diets on the turnover rates of servan, liver and muscle proteins in the rat, measured by continuous infusion of L-(U-^{14}C)- lysine. Clin. Sci., 35, 287-305.

Wiggins, J.P., Rothenbacher, H., Wilson, L.L., Martin, R.J., Wangness, P.J. and Ziegler, J.H. 1980. Growth and endocrine responses of lambs to zeranol implants. J. Anim. Sci., 49, 291.

DISCUSSION ON DR. BUTTERY'S PAPER

Dr. Galbraith: What is the effect of the various anabolic agents on dietary intake of treated animals?

Dr. Buttery: Generally the tendency is for increased dietary intake but we did not get much of an increase in that particular trial with sheep. If they do stimulate food intake then you would expect them to stimulate protein synthesis in the muscle, not to depress it.

Dr. Heitzman: Most of those effects on growth can be achieved with a fixed food intake and so there is still a feed conversion problem.

Prof. Hoffmann: Trenbolone has been considered as a so called androgen. In many papers it has been directly compared to testosterone. We have found a difference in the distribution of the residues of trenbolone acetate and testosterone. Testosterone residues were highest in the kidney while trenbolone residues were highest in the liver. In your work have you any direct comparison between trenbolone acetate and testosterone rather than durabolin?

Dr. Buttery: We are currently giving testosterone to female rats and looking at the effects on muscle metabolism and we may have the answer to your question shortly. We have some evidence that suggests that testosterone is working by stimulating protein synthesis. Interestingly, we have done the classical levator ani tests and trenbolone appears more androgenic than testosterone.

Dr. Heitzman: Trenbolone and zeranol may be active in the same way in depressing protein synthesis whereas testosterone presumably increases protein synthesis and the combined preparations of testosterone and oestrogens are very effective. Can you comment on that?

Dr. Buttery: There is evidence that trenbolone is affecting muscle degradation rates more than synthetic rates. It is possible that testosterone is working more on synthesis. We did once inject trenbolone and durabolin into the same animal and got growth stimulation which was similar to either trenbolone or durabolin alone. There were no indications of an additive effect.

Dr. Roche: How do you explain the effect of trenbolone on increasing so called oestradiol levels in female sheep?

Dr. Buttery: One must be careful about this. It could be that we got a metabolite that was binding to the oestradiol antibody. We did not check this. I don't have an explanation for these observations.

Dr. Heitzman: We found the same situation in mature female cattle but this is because of an interference with oestrous behaviour and maintaining the animals in a phase in which they are secreting higher oestrogens.

Dr. Roche: So it may well be an effect on the follicular population on the ovaries inducing cysts.

Prof. Hoffmann: What was the dose you gave to the sheep and did you measure liver function?

Dr. Buttery: 80mg. of trenbolone which is higher than would be commercially recommended. We did not do liver function tests.

Prof. Hoffmann: You might just have affected liver function.

Dr. Meyer: I would like to comment on the oestradiol receptor which you found in muscle. Are you certain that this was an oestradiol receptor because you did your scatchard

plots in a very unusual range. The plot was in the range
where you had only one point to make the slope and this
was in the range where the free was present in a 20 fold
excess. The number of binding sites reported was
25 - 40% of those found in the uterus. This seems very
high so I wonder if what you are measuring are oestradiol
receptors?

Dr. Buttery: We have tried displacing oestradiol with
various steroids and the binding does seem to be specific
for oestradiol but I accept you criticisms as they are only
very preliminary observations. But other people have
shown that oestrogens receptors do occur in muscle,
particularly in rats.

Dr. Galbraith: We have been investigating the relative
binding behaviour of certain steroids in cytosol preparations
from liver and hind limb muscle of castrate male sheep.
We have observed in certain of these studies that the
specific binding of $[^3H]$-dexamethasone or $[^3H]$-cortisol is
reduced in the presence of trenbolone and testosterone.
These results may support previous suggestions that
protein accretion in muscle may be increased by interference
with catabolism associated with the activity of glucocorticoids.

Prof. Karg: You indicated in your talk a multi-component
mode of action that oestrogens promote GH secretion in rats.
This is a dose problem rather than a species specific
problem. Release of prolactin by oestrogens occurs in a
dose response fashion, and it goes from positive to negative.
Another example concerning the catabolic action of glucocorticoids
is that it is a dose related effect. If you take a very low
dose you always have an anabolic action mainly connected with
other effects like appetite increase etc.

Dr. Buttery: I do believe there is a multi-component mode
of action. There may be one dominant factor but there are
several things happening.

Dr. Galbraith: The possibility exists that trenbolone acts in an oestrogen like fashion as a general effect. You mentioned that oestrogens inhibit the growth of rodent animals, particularly rats. I find it difficult to understand how trenbolone could stimulate growth in rats through an oestrogen effect when it is well known that oestrogens, at certain dose levels at least, have the opposite effect in this type of animal.

Dr. Buttery: It has been suggested that very low levels of oestrogens can promote growth in rats.

Prof. Karg: I agree that very low doses (0.05 μg/day) of oestrogens in rats have an anabolic effect.

EFFECT OF ANABOLIC STEROIDS AND GROWTH HORMONE
ON MAMMARY DEVELOPMENT IN HEIFERS

K. Sejrsen

National Institute of Animal Science
Rolighedsvej 25, DK-1958 Copenhagen,
Denmark

ABSTRACT

The limited data available on the effect of anabolic steroids on mammary development during growth of heifers indicate that oestrogens may have stimulating effect on duct growth, but may cause abnormal alveolar development, when given to heifers after puberty. Progestins may have a stimulatory effect, but the effect does not seem to be cumulative with normal growth of the mammary gland during pregnancy. Androgens - at least trenbolone acetate - are detrimental to mammary growth and should not be used for heifers. The effect of exogenous growth hormone on mammary growth in heifers has not been investigated, but the available information suggests that use of exogenous growth hormone may stimulate mammary growth and this effect may very well be of greater importance than the growth stimulating effect in heifers.

INTRODUCTION

Increased feed efficiency of heifers during growth is of obvious benefit to dairy as well as beef industry. Anabolic steroids increase growth rate and feed conversion ratio in growing cattle (Heitzman, 1976) and the growth stimulating effect of growth hormone has long been recognized. Exogenous growth hormone has been shown to stimulate growth in dairy heifers (Brumby, 1959). Use of exogenous growth hormone has become a practical possibility, since bovine growth hormone has been produced by the recombinant DNA technique and recombinant bovine growth hormone has been shown to stimulate milk yield similar to natural bovine growth hormone (Bauman et al, 1982).

Before anabolic agents are recommended for use for replacement heifers it is necessary to consider, whether lactational - and reproductive - performance is affected. Mammary development must be unimpaired, since the number of

milk secreting cells is one of the primary determinants of
milk production (Tucker, 1981).

NORMAL MAMMARY DEVELOPMENT
     The fully developed mammary epithelium consists of
branching ducts and lobes and lobules of alveoli (Cowie &
Tindal, 1971). There is no alveolar development before pregnancy, and most alveoli are developed during the last half
of pregnancy. It is therefore mainly ductular development
that has to be considered before anabolic agents are used.
     The gland anlage exists during fetal life but the ducts
are rudimentary at birth. The period after birth is a
quiesent phase and there is no ductular growth. In this
period the glands grow at the same rate as the body as a
whole - the growth is isometric. At about 3 months of age
there is an increase in the growth rate of the mammary
glands compared to the body - the growth becomes allometric
(Sinha & Tucker, 1969). There is a rapid increase in the
size of the fat pad and the ducts grow into the fat pad. The
period of allometric growth continues until puberty, when
growth of the ducts gradually becomes isometric. The isometric period of development continues until pregnancy,
where lobulo-alveolar development takes place.

EFFECT OF ANABOLIC STEROIDS ON DUCTULAR DEVELOPMENT
     The ducts develop into the fat pad and the fat pad
therefore predetermines the maximal outgrowth of the ducts
(Faulkin & De Ome, 1960, Mayer & Klein, 1961). However, in
order to grow the ducts require a hormonal stimulus and the
onset of allometric growth coincide with the onset of
oestrogen secretion from the ovary in the rat (Cowie, 1949)
and the heifer (Wallace, 1953). The role of oestrogen secretion has long been recognized (Turner, 1939, cited from
Cowie & Tindal, 1971) and Wallace (1953) observed that administration of the synthetic oestrogen diethyl stilbestrol
(DES) to calves caused increased spread of the ductular
epithelium. Synthetic oestrogen given to non pregnant heifers

after puberty has resulted in lobulo-alveolar development
and induction of lactation, when the heifers were milked
(Cowie & Tindal, 1971). The alveolar development, however,
included several abnormalities and deficiency of secretory
cells. The alveolar development may not have been due to
oestrogen, since the milking stimulus alone can induce
lactation, most likely via stimulation of the release of
pituitary hormones. Nevertheless care should be taken
before oestrogens are generally recommended for use in
heifers, especially after puberty.

Progesterone is only secreted in very low amounts
prior to puberty and progesterone is not required for
ductular development (Lyons et al, 1958, Nandi, 1959).
Tucker (1981) suggests that progesterone is involved in
the failure of maintaining allometric growth after
puberty via the asynchronous secretion with oestrogen
during oestrus cycles. During pregnancy when progesterone
and oestrogen are elevated simultaneously lobular-alveolar
development is stimulated and injection of progesterone
in combination with oestrogen causes lobulo-alveolar
growth and induces lactation in virgin heifers (Tucker,
1981). Pritchard et al (1972) gave the synthetic pro-
gestin melengestrol acetate (MGA) to Holstein heifers
from 2.5 months of age to 1st oestrus and to 120 cm withers
height - approximately 350 kg liveweight (Table 1). The

TABLE 1 Mammary parenchyma and DNA of left udder halves
at 120 cm withers height and milk yield of
heifers given 45 mg melengestrol acetate (MGA)
orally per day from 2.5 months of age to either
1st oestrus or 120 cm withers height, (Pritchard,
1970, Pritchard et al, 1972)

	Control	MGA from 2.5 months to 1st oestrus	120 cm
Mammary parenchyma, g	599	788	736
Mammary DNA	1501	2564*	2369*
Milk yield	979	931	1004

*Significantly different from control.

heifers receiving MGA to 1st oestrus and 120 cm had 32 and
23% more mammary parenchyma and 70 and 58% more mammary
DNA respectively than control animals. The increased mammary growth, however, did not seem to be cumulative with normal growth, since there was no difference between treatments
in milk yield during the first 60 days of 1st lactation.

Androgens obviously are not involved in regulation of
normal ductular development in heifers. Little et al (1979)
investigated the effect of subcutaneous implantation of
300 mg trenbolone acetate (T) and 140 mg T + 20 mg oestradiol at 16 and 31 weeks of age on milk production in British
Friesian heifers (Table 2). The milk yield of the heifers
receiving 300 mg T was only 18% of the controls. The milk
yield of the second treatment group was intermediate. The
authors suggested that this was due to the lower dose of T,
but another possibility is a stimulatory effect of the
oestradiol. Mammary development was not measured, but it
was concluded that impaired mammary development was the
main cause for the lower milk production.

TABLE 2   Milk yield and milk composition of heifers
          implanted with 300 mg trenbolone acetate (T)
          or 140 mg trenbolone acetate + 20 mg oestradiol
          (TE) at 16 and 31 weeks of age. (Little et al,
          1979)

	Control	T	TE
Milk, kg	3636	641*	1992*
Fat%	3.81	4.95*	3.79
Protein%	3.12	3.43*	3.25
Lactose%	4.66	4.56	4.56
4% fat corrected milk, kg	3563	652*	1911*
Length of 1st lactation, days	293	96*	212

* Significantly different from control.

The limited data available on the effect of anabolic steroids on mammary development indicate that oestrogens may have stimulatory effects on duct growth, but may cause abnormal alveolar growth after puberty. Progestins may have a stimulatory effect, but the effect does not seem to last through pregnancy. Androgens - at least trenbolone acetate - are detrimental to mammary growth and should not be used for replacement heifers.

EFFECT OF EXOGENOUS GROWTH HORMONE ?

The pituitary hormones appear to play a major role in stimulating mammary development, since these hormones alone are capable of stimulating mammary development in rats (Talwalker & Meites, 1961) and goats (Cowie et al, 1966), whereas the ovarian hormones have no mammogenic activity in hypophysectomized animals.

The minimal hormonal requirement for ductular development in rats and mice has been established in extensive replacement therapy studies by Lyons and coworkers (Lyons et al, 1958) and Nandi (1959) respectively. They demonstrated that growth hormone in both species is required for ductular development and Lyons et al (1958) demonstrated a direct effect of growth hormone on the mammary parenchyma. Even if it was stressed that species differences may exist Cowie et al (1966) found the same hormonal requirement for lobulo-alveolar development in goats as was found in rats and mice. Cowie et al (1966) did not investigate ductular development, but the results on alveolar development suggest a similar hormonal requirement in ruminants as in rodents. It is therefore likely that growth hormone also in heifers is of major importance for ductular development.

Even if growth hormone is required for ductular development, growth hormone may not be rate limiting. We, however, have observed a positive association and linear re-

gression between circulating levels of growth hormone and
ductular development during the allometric phase in heifers
fed on two levels of feeding (Table 3 and 4) (Sejrsen et
al, 1982, a, b).

TABLE 3   Daily gain, mammary parenchymal and adipose
tissue and mean serum concentrations of growth
hormone in heifers from 175 to 320 kg and from
300 to 440 kg live weight on two feeding levels
(Sejrsen et al, 1982a, b)

	Body weight Range, kg	Feeding level Normal	High
Daily gain, g	175-320	$637^{\pm}72^{b}$	$1271^{\pm}66$
	300-440	$588^{\pm}72^{b}$	$1164^{\pm}66$
Mammary parenchyma, g	175-320	$642^{\pm}65^{d}$	$495^{\pm}60$
	300-440	$987^{\pm}110$	$957^{\pm}100$
" " , %	175-320	$38^{\pm}1^{b}$	$23^{\pm}1$
	300-440	$37^{\pm}1$	$32^{\pm}1$
Mammary adipose, g	175-320	$1040^{\pm}125^{b}$	$1708^{\pm}113$
	300-440	$1751^{\pm}297$	$2113^{\pm}271$
Growth hormone,ng/ml[a]	175-320	$5,3^{\pm}0,8^{c}$	$3,7^{\pm}0,8$
	300-440	$3,3^{\pm}0,6$	$3,1^{\pm}0,5$
No. of animals	175-320	5	. 6
	300-440	5	6

[a] Mean of 104 samples per animal at 30 minutes interval at two 24 hour periods.

b) P < 0,01,   c) P < 0,10,   d) P < 0,13

TABLE 4   Relationship between measures of mammary gland
          growth and serum concentrations of growth
          hormone (Sejrsen et al, 1982b)

	Body weight range, kg	Correlation	Regression[a]
Mammary parenchyma,g	175-320	0,45	255[b]
	300-440	0,19	51
Mammary parenchyma,%	175-320	0,56[d]	13,2[b]
	300-440	0,55[d]	3,2
Mammary adipose, g	175-320	-0,53[d]	-542[c]
	300-440	-0,57[d]	-105
Mammary adipose, %	175-320	-0,56[d]	-13,2[b]
	300-440	-0,55[d]	- 3,2

a) Regression coefficient after adjustment for serum
   concentrations of prolactin, insulin and glucocorticoids in a multiple regression analysis.

b) P < 0,01

c) P < 0,05

d) P < 0,10.

These results obviously do not prove a cause and effect
relationship between circulating levels of growth hormone
and duct growth, but a cause - effect relationship may very
well exist, especially if growth hormone has a direct
effect on mammary parenchyma in heifers as in rats. If this
is the case, the data strongly suggest that the negative
effect of high feeding level on mammary development and
subsequent milk yield, is caused by the lower circulating
growth hormone concentration in heifers on high feeding
level. This would similarly suggest that use of exogenous
growth hormone in heifers stimulates mammary development,
and this effect might very well be of greater importance
than the growth stimulating effect.

REFERENCES

Baumann, D.E., M.J.De Geeter, C.J.Peel, G.M.Lanza,
    R.C.Gorewit & R.W.Hammond. 1982. Effect of recombi-
    nantly derived bovine growth hormone (bGH) on lacta-
    tional performance of high yielding dairy cows. J.
    Dairy Sci. 65: Suppl. 1. P.86.
Brumby, P.J.1959. The influence of growth hormone on growth
    in young cattle, New Zealand. J.Agr.Res. 2:683-689.
Cowie, A.T. 1949. The relative growth of the mammary growth
    of the mammary gland in normal, gonadectomized and
    adrenalectomized rats. J.Endocr.6, 145-157.
Cowie, A.T., J.S.Tindal & A.Yokoyama. 1966. The induction
    of mammary growth in hypophysectomized goat. J.Endocr.
    34, 85.
Cowie, A.T. & J.S.Tindal.1971. The physiology of lactation.
    Butterworth,London. 392 pp.
Faulkin, L.J. & K.B.DeOme. 1960. Regulation of growth and
    spacing of gland elements in the mammary fat pad of
    the C3H mouse.. J.Natl. Cancer.Inst.24, 953.
Heitzman, R.J. 1976. The effectiveness of anabolic agents
    in increasing rate of growth in farm animals; report
    on experiments in cattle. In: Anabolic Agents in
    Animal Production, p. 89-98. Ed.F.C.Lu & J.Rendel,
    Georg Thieme, Stuttgart.
Little, W., R.M.Kay, D.J.Harwood & R.J.Heitzman.1979.
    The effects of implanting prepuberal dairy heifers
    with anabolic steroids on live weight gain, blood and
    urine composition and milk yield. J.Agric.Sci.,Camb.
    93, 321-327.
Lyons, W.R., C.H.Li & R.E.Johnson. 1958. The hormonal
    control of mammary growth. Recent.Prog.Horm.Res.14,219.
Mayer, G. & M.Klein. 1961. Histology and cytology of the
    mammary gland. In: Milk: The mammary gland and its
    secretion. Ed. S.K.Kon & A.T.Cowie. A.P.London.
Nandi, S. 1959. Hormonal control of mammogenesis and lacto-
    genesis in the C3H/He Crgl. mouse. Univ.Calif.Publ.
    Zool 65, 1.
Pritchard, D.E. 1970. Endocrine and reproductive changes in
    dairy heifers as affected by growth rate and melenge-
    strol acetate. Ph.D.thesis. Michigan State University,
    East Lansing.
Pritchard, D.E., H.D.Hafs, H.A.Tucker, L.J.Boyd, R.W.Purchas
    & J.T.Huber. 1972. Growth mammary, reproductive and
    pituitary hormone characteristics of Holstein heifers
    fed extra grain and melengestrol acetate. J.Dairy Sci.
    53, 995.
Sejrsen, K., J.T.Huber, H.A.Tucker & R.M.Akers. 1982. In-
    fluence of nutrition on mammary development in pre-
    and postpubertal heifers. J.Dairy Sci. 65, 793.
Sejrsen, K., J.T.Huber & H.A.Tucker. 1982. Influence of feed-
    ing level on hormone concentrations and their relation-
    ship to mammary growth in heifers. Submitted to J.Dairy
    Sci.

Sinha, Y.N. & H.A.Tucker. 1969. Mammary development and
      pituitary prolactin level of heifers from birth
      through puberty and during the estrous cycle. J.Dairy
      Sci. 52, 507.
Talwalker, P.K. & J.Meites. 1961. Mammary lobulo-alveolar
      growth induced by anterior pituitary hormones in adre-
      no-ovariectomized and adreno-ovariectomized hypophys-
      ectomized rats. Proc. Soc. Exp. Biol.Med. 107, 880.
Tucker, H.A. 1981. Physiological control of mammary growth,
      lactogenesis and lactation. J.Dairy Sci. 64, 1403.

242

DISCUSSION ON DR. SEJRSEN'S PAPER

Dr. Roche: In relation to the effect of plane of nutrition on mammary gland development, an area of relevance would be the beef suckler replacement heifer. These animals are most likely to be on a higher plane of nutrition than 600g. per day during the pre-puberal period. Is it possible that this high plane of nutrition may be one reason for lower milk yield in beef cows?

Dr. Sejrsen: I think this is very likely. I think it is well demonstrated that heifers raised in good years become poor cows as regards milk produciton.

Dr. Hetizman: In the beef cow one refers to the flip-flop situation. A heifer well fed early in life, that subsequently calves has a poor milk produciton, and its calf therefore grows slowly and has a good milk production and so on. This is a recognised problem. In our work at Compton we have measured the impairment of mammary gland development in animals of different weights. We find that in animals grown at very high rates, where we get considerable impairment of lactational performance you get low growth hormone levels and high insulin concentrations in plasma.

Dr. Forbes: You mentioned that you were doing work with exogenous growth hormone. Could you comment on the results?

Dr. Sejrsen: We are working with identical twins.

Dr. Forbes: You start the injections at quite a young age.

Dr. Sejrsen: We start at about 175kg. and continue for 100 days and try to have about 800 grams of daily gain.

Dr. Forbes: Are you getting a positive response in mammary development?

Dr. Sejrsen: We do not know yet as we only have one pair of twins but there is an increase in growth rate.

Dr. Galbraith: Do you have any thoughts on the role of prolactin in the processes you have been describing?

Dr. Sejrsen: Yes, because we measured prolactin, insulin and glucocorticoids. Prolactin was significantly higher in animals on a high plane of nutrition, and there was a significant negative relationship between mammary development and prolactin. That makes' one wonder what photoperiod does to mammary development? I did multiple regression analysis to try to find out whether it is the change in growth hormone or prolactin that causes the decrease in mammary development in heifers raised on a high feeding level. After adjustment there was no relationship between prolactin and mammary development but the relationship with GH was maintained.

Prof. Karg: You mentioned that the effect of prolactin might be due to a seasonal effect. Could you speculate on the effect of trenbolone on inhibiting the development of the mammary gland?

Dr. Sejrsen: With regard to prolactin and seasonality, all animals had 16 hours light and the temperature was recorded. As regards the effect of trenbolone, I will have to refer to Dr. Heitzman.

Dr. Heitzman: We have found seasonal effects on the development of the mammary gland in rapidly reared heifers. There is a difference because the age of puberty is different because of season. We think puberty is the critical point which terminates this first allimetric phase of growth in the mammary gland. Therefore season can have an effect. With regard to trenbolone we have measured mammary glands. It seems to be purely an androgenic effect. I do not know what that means except that there seems to be considerable inhibition of any mammary gland development if

244

the heifer is exposed to large circulating levels of trenbolone in the prepubertal period. The mammary gland is almost
all fat, very little ductal tissue, and very little
parenchymal infiltration of the fat. Milk yields are very
low. You also see other problems associated with the
androgenic effects in this prepubertal period. You get
dystocia, and virilization of the genetalia.

Prof. Karg: The pre-pubertal period is not the time of sexual
differentiation but we know the sensitivity of the hypothalamic
releasing centres could be affected. One of the effects of
trenbolone could be on the sensitivity of releasing centres.

Dr. Schanbacher: In work we have done we noticed that the
central nervous system and particular parts of it are sensitive to both androgens and oestrogens. We have observed that
both testosterone and oestradiol and climate will increase
prolactin. So I don't know whether prolactin is stimulatory
to the mammary gland but relative to the effects that I
thought prolactin had, it is certainly interesting. The
central effect of the steroids on brain function is hard to
interpret. We are trying to sort out androgenic and
oestrogenic effects, and then thinking of the synthetic
compounds like trenbolone certainly provide for an interesting
subject. We should not neglect prolactin because administration of prolactin inhibitors decrease mammary gland
development.

Prof. Karg: We have shown this in the prepubertal period.

Dr. Sejrsen: It is also a question of what hormones are
required in a minimal dose and it is necessary to have a
sufficient amount of prolactin.

Prof. Karg: Concerning the effect of steroids on the relase of
prolactin, it is very well documented that very low doses are
stimulatory and high doses are inhibitory. This is in
contrast to the well known lactation curve during pregnancy.

Dr. Heitzman: In your discussion you suggest that it is the
progesterone in the first oestrous cycle that might terminate
the development of this first phase. There is also the
possibility that it may be some changes in the responsiveness
of the pituitary at the onset of puberty which might bring
about these effects. Would you like to comment?

Dr. Sejrsen: When you consider the negative effect of
plane of nutrition on mammary development, you can think
of different possible causes. One is the view which you hold
at the moment related to when onset of puberty takes place.
With increased plane of nutrition puberty occurs earlier and
there is a shorter period of allimetric growth. This means
less mammary growth if the growth rate is unchanged.
However, another possibility is the one that I suggest, namely
that growth rate of the mammary gland during the allimetric
growth period is decreased. The two possibilities
however, do not exclude each other and may in reality
go hand in hand.

PHOTOPERIODISM AND HORMONES IN SHEEP AND GOATS

M. Terqui*, C. Delouis**, R. Ortavant*

INRA - Département de Physiologie Animale
*Station de Physiologie de la Reproduction
Nouzilly, 37380 MONNAIE, FRANCE
**Laboratoire de Physiologie de la Lactation
78350 JOUY EN JOSAS, FRANCE

ABSTRACT

In the adult ram, experiments with a flash of light during the dark phase (scotophase scan experiments) demonstrate that, for LH, T, FSH, prolactin and testis diameter, photoperiodic time is measured by a rhythm of photosensitivity. A flash of one hour duration gave maximal stimulation (photosensitive phase) for LH, T, FSH and testis when the flash occurred between 16-17 hours after dawn; for prolactin maximal stimulation occurred when the flash was given 9 to 10 hours after dusk.

In goats induced to lactate by steroid treatments, photoperiodic regime modify prolactin and GH secretion. Long daylength increased prolactin and also GH levels compared to levels during short daylength.

Thus photoperiodic treatment effects on reproduction and lactation involve modification of secretion of LH, T, FSH, prolactin and GH. For the first four hormones a photosensitive phase exist which may not be controlled by the same synchroniser - dawn or dusk.

INTRODUCTION

Reproductive activity of many breeds of domestic sheep and goats is restricted to certain months of the year (Thibault et al., 1966). Among the many environmental factors in our temperate climate, photoperiodism is the main factor responsible for the seasonal breeding of sheep. The effects of various treatments have been investigated both in male and female sheep (Ortavant, 1977) such as a "reverse" annual variation in daylength (Alberio, 1976). or six month rhythm (Ortavant and Thibault, 1956; Mauleon and Rougeot, 1962).

From these experiments and others arise at least two questions:

i) How photoperiodic time is measured by an animal?

ii) Is this time measurement the same for all hormones in
a breed?

Two groups of hypothesis have been proposed to explain
photoperiodic measurement.

In the first, time measurement is based on some form of
hourglass or interval timer. Those in favor of this
hypothesis propose a critical duration of light or dark or
a critical ratio of light and dark in each cycle.

The second was proposed by Bünning in 1936 and is based
on an endogenous circadian rhythm of photosensitivity. In
such a model, the photoperiodic effect of a particular light:
dark cycle depends on the position of light relative to the
circadian rhythm of sensitivity. Figure 1 shows one protocol
to demonstrate the existence of the photosensitive phase, i.e.
the so called scotophase scan experiment.

In this paper, the effects of such a protocol on
endocrine secretions and testis growth and those of short
and long daylength on lactating goats are reported

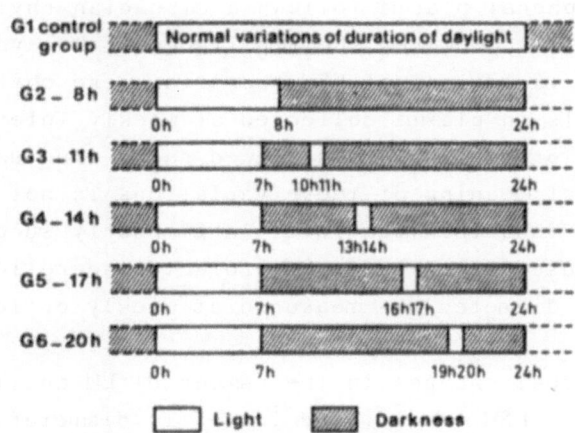

Figure 1 : Experimental protocol to demonstrate a photosensitive phase.

PHOTOSENSITIVE PHASES IN RAM FOR LH, TESTOSTERONE, FSH, PROLACTIN AND TESTIS

I. Methodology

The scotophase scan experiments (Figure 1) have been used to establish the existence of a photosensitive phase. The abrupt beginning of the main light phase is called dawn and the end dusk. The dark period is interrupted by a flash of light of 1 hour which is given at various intervals after the main light phase. When the flash occurs 7 to 8 h after dawn, the rams are submitted to 8 h of continuous light. The light is given by fluorescent tubes providing 300 lux at ram eyes. Adult rams were used in all the reported experiments.

It is known that LH and testosterone (T) are secreted as pulses. The number of LH and T pulses during 24 hours is the important parameter from the physiological point of view (Terqui et al., 1980). Furthermore, these pulses do not occur at random (Terqui et al., 1980). Therefore, the sampling strategy was to collect jugular blood every hour for 24 hours at various times during photoperiodic treatment. FSH in peripheral plasma follows a circadian rhythm (Blanc et al., 1981). However the amplitude of this variation is low, so the variations of FSH secretion were characterized by the levels in plasma collected at weekly intervals.

Plasma prolactin levels showed rapid fluctuations. The physiological meaning of these variations is not yet clearly understood; thus the mean level in 24 hourly successive samples was used as an index of prolactin secretion.

Testis diameter was measured at weekly or fortnightly intervals.

Individual changes in the number of LH and T pulses, in prolactin and FSH levels, and in testis diameter were analyzed by non parametric tests which are more relevant for small number of animals per group (5 to 6), for non independent data with heterogenous and correlated variance.

II. <u>Photosensitive phase for LH, testosterone, FSH and testis
      diameter</u> (Pelletier <u>et al</u>., 1981).

Since the photoperiodic regime before the beginning of
any experiment on photoperiodism might be extremely important,
two sorts of experiments have been done with animals pre-
treated with either long days (Experiment A) or with short
days (Experiments B and C).

<u>Experiment A</u> was begun in June and before the experiment,
the rams were subjected to normal daylength.. Therefore , they
were long-day pre-treated animals. There were six groups of
rams: one group continued on normal variations in daylength
(G1), while the other five received a main light phase of
seven hours plus a one hour flash starting either 7 hours
(G2), 10 hours (G3), 13 hours (G4), 16 hours (G5) or 19 hours
(G6) after the beginning of the main light phase.

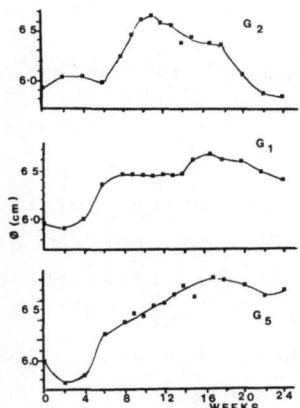

<u>Figure 2</u>: Mean testis diameter variations in 8-hour group
            (G2) in normal light group (G1) and in 16 hour
            group (G5).

Figure 2 shows the variation in testis diameter for 3 of
the groups of rams. For the two flash treated group, testis

size increased a few weeks after the beginning of treatment.
However, only in the group with a flash starting 16 hours
after dawn, testis diameter variations were similar to those
of the normal daylight group. Large testis were maintained
for several weeks. In the three other groups (flashes at 10,
13 or 19 hours), the pattern of variation in testis size was
very close to that of group G2 i.e. an increase of short
duration followed by a decrease.

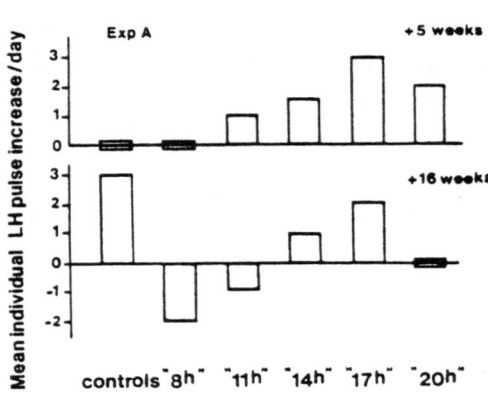

Figure 3: Variation of the increase of the mean number of
LH pulses in 24 hours per ram at 5 weeks (upper)
and 16 weeks (lower) after the beginning of
photoperiodic treatment.

LH and T pulse numbers were also affected by the position
of light flash as shown in Figure 3.

There was a clear effect of the timing of the flash on
the increase in LH pulses, the greatest being for the 16-hour
group (G5) after 5 weeks of photoperiodic treatment. Even at
16 weeks G5 still had the greatest increase in pulses
relative to the other treated groups.

Experiment B was begun in January, and before the experiment,
the rams were subjected to normal daylength. Therefore, they

were short-day pre-treated animals. There were six groups of
rams treated as for Experiment A.

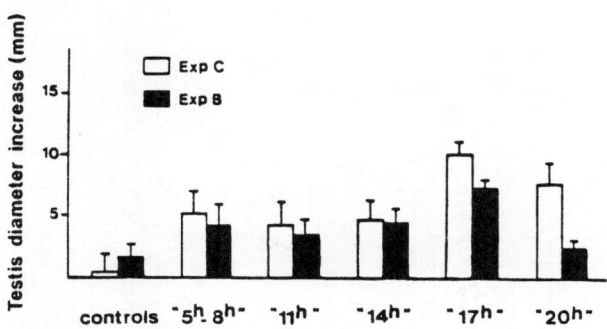

Figure 4:   Increase in mean testis diameter in Experiment B
             (short-day pre-treated rams with main light phase
             of 7 hours) and Experiment C (short-day pre-treated
             rams with main light phase of 4 hours).

       As presented in Figure 4, the testis diameter increase
was greatest in the group which received a flash 16 hours
after dawn. Accordingly, whatever the pre-treatment short
or long days, a flash of one hour given 16 hours after dawn
stimulated LH, T and FSH secretion.

Experiment C was carried out to determine if either dawn or
dusk was the synchroniser of gonadotrophic activity. The
rams were pre-treated with short-days. There were six groups
treated as for Experiments A and B except that the duration
of the light phase was only 4 hours, and a seventh group was
added which received a one hour flash starting 4 hours after
dawn.
       The results for testis diameter in Figure 4 show that,
as for previous experiments, maximal stimulation occurred
in the group flashed at 16 hours.

Figure 5 summarizes all the results for LH, T, FSH and testis diameter. This figure also indicates that in natural conditions testis growth is initiated during long days; this is relevant to the demonstration of the same photosensitive phase 16 hours after dawn for LH, T, testis and probably al for FSH.

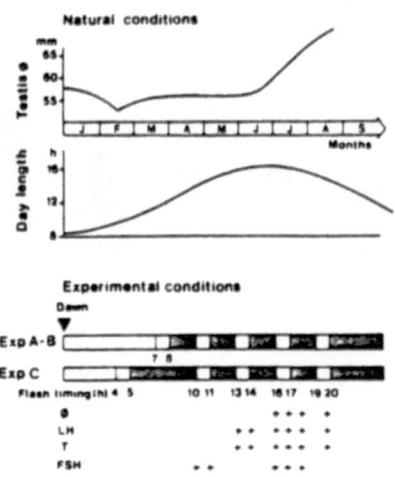

Figure 5: LH, T, FSH, testis diameter photosensitive phase and natural variation of daylight and testis size

PHOTOSENSITIVE PHASE FOR PROLACTIN (Revault et al., 1981).

Prolactin is extremely sensitive to photoperiod and changes in mean levels are very clear.

in Experiments A and B, with 7 hours of main light phase, whatever the pre-treatment (short or long days) the maximum stimulation occurred when a one hour flash was given 16 hours after dawn.

However, in Experiment C, with a main light phase of 4 hours the photosensitive phase occurred 13 to 14 hours after dawn.

In order to clarify this a fourth experiment was performed with a main light phase of 10 hours. With this treatment, the photosensitive phase moved 19-20 hours after dawn.

These results are summarized in Figure 6, and it is obvious that dawn is not the synchroniser but it is dusk and that the prolactin photosensitive phase is between 9 to 10 hours after dusk.

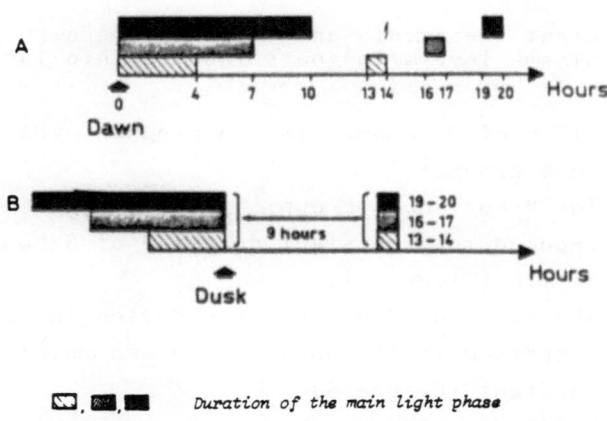

Figure 6: Photosensitive phase for prolactin position of those light pulses which were stimulatory for maximal prolactin secretion in rams.
A   dawn as origin
B   dusk as origin.

EFFECT OF EXTENDED DAYLENGTH IN GOATS ON PROLACTIN, GROWTH HORMONE AND LACTATION (Delouis et al., 1983)

Lactation is controlled by many hormones of maternal and foetoplacental origin (Delouis et al., 1980). A more simple model is the non pregnant goat induced to lactate by treatment with oestrogen, progesterone and corticoids.

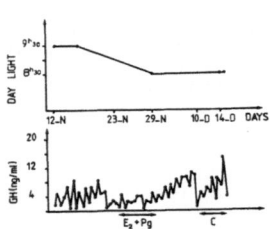
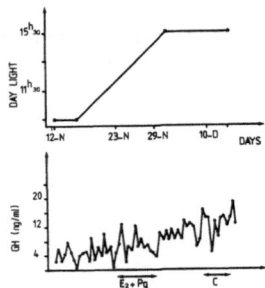

Figure 7: Light treatments and mean daily growth hormone (GH) plasma levels in goats induced into lactation.

On the 12th of November, 18 non pregnant goats were allotted into 2 groups:

    i)   The "short-day" group: the daylength was reduced to a constant duration of 8 hours 30 min (Figure 6).

    ii)   The "long-day" group: the daylength was increased to 15 hours 30 min and maintained constant (Figure 6).

From the 23rd to 29th of November, oestradiol 17β (0.5mg/ kg/day) and progesterone (1.25mg/kg/day) were administered subcutaneously to each female of both groups. Ten days later cortisol (25mg/day) was injected for five days. Milking began during cortisol treatment (Figure 7).

Mean daily plasma levels of growth hormone varied during treatment in both groups, but the levels were higher in the long-day group than in the short-day group.

Since growth hormone secretion is pulsatile the 24 hours variation of this hormone was studied in peripheral plasma collected hourly before milking was begun and the differences between the two groups are more clear (Figure 8).

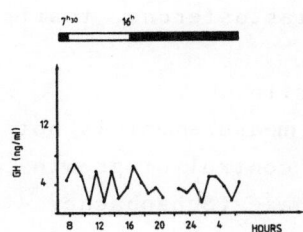
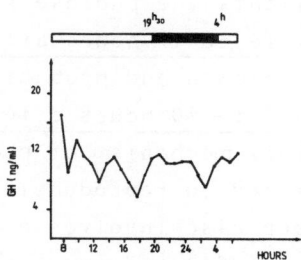

Figure 8: Mean hourly growth hormone (GH) levels in plasma in "short-day" and "long-day" groups. The GH plasma levels are higher in long day group than in short day group.

Prolactin levels are also markedly influenced by light regime and they are much higher in long day group than in short-day group.

Thus, growth hormone and prolactin secretion were increased when the duration of daylight was extended to 15 hours 30 min. As a consequence of these changes of both hormones and probably others, milk productions between the two groups differed.

The mean adjusted milk production of goats subjected to long daylength was higher than the production of the short day group ($P<0.05$). The milk composition also seemed modified. Similar effects on milk production have also been observed but to a lesser extent in the cow (Peters et al., 1981).

Long days appear to modify hormone secretion, such as growth hormone and prolactin, after a steroid treatment to induce lactation. This suggests that photoperiodism or season of the year might have an important effect on the response to other steroid treatment such as anabolic steroids.

CONCLUSIONS

Sheep measure photoperiodic time by an endogenous circadian rhythm of photosensitivity. From scotophase scan experiments the photosensitive phase occurs:
- 16 to 17 hours after dawn for LH, testosterone, testis growth and probably FSH,
- 9 to 10 hours after dusk for prolactin.

This mechanism of photoperiodic time measurement is not restricted to reproduction; photoperiodic control of growth in sheep also involves a photosensitive phase (Schanbacher et al., 1981) and our experiments on lactation show that GH secretion, at least partially, is under photoperiodic control

The data available suggest that the photosensitive phase may not be controlled by the same synchroniser - dawn or dusk.

ACKNOWLEDGMENTS

The authors are grateful to Dr. B. KILGOUR for suggestion and help in the preparation of the English manuscript.

REFERENCES

Alberio, R. 1976. Rôle de la photopériode dans le developpement de la fonction de reproduction chez l'agneau Ile-de-France de la naissance à 21 mois. Thèse Doc. 3è Cycle, Univ. Paris VI, pp 57.

Blanc, M.R., Daveau, A., Ortavant, R., Pelletier, J., Revault J.P., and de Reviers, M.M. 1981. Circadian variations of LH, FSH and Prl levels in plasma of Ile-de-France or Prealpes du sud rams in photoperiodism and reproduction. (ed. R. Ortavant, J. Pelletier and J.P. Revault). (INRA publ.), pp 99-115.

Bunning, E. 1936. Die endogene Tagesrythmik als Grundlage der photoperiodischen Reaktion. Ber Deat. Bot. Ges. 54, 590-607.

Delouis, C., Djiane, J., Houdebine, L.M. and Terqui, M. 1980. Relation between hormones and mammary gland function. J. Dairy Sci., 63, 1492-1513.

Delouis, C., Kahn, G., Hart, I.C., Fevre, J. and Ortavant, R. 1983. The effect of daylength duration on prolactin, growth hormone and corticoids in goats induced in lactation (to be published).

Mauleon, P. and Rougeot, J. 1962. Régulation des saisons
     sexuelles chez des brebis de races différentes au moyen
     de divers rythmes lumineux. Ann. Biol. Anim. Bioch.
     Biophys., 2, 209-222.

Ortavant, R. and Thibault, C. 1956. Influence de la durée
     quotidienne d'éclairement sur la spermatogenèse du
     Bélier. Proceed. IInd. Cong. Fert. Steril. 12-43.

Pelletier, J., Blanc, M., Daveau, A., Garnier, D.H., Ortavant,
     R., de Reviers, M.M. and Terqui, M. 1981. Mechanism of
     light action in the ram. A photosensitive phase for
     LH, FSH, testosterone and testis weight? In: "Photo-
     periodism and Reproduction". (Ed. R. Ortavant, J.
     Pelletier and J. P. Ravault) (INRA Publ). pp 117-134.

Peters, R.R., Chapin, L.T., Emery, R.S. and Tucker H.A. 1981.
     Milk yield, feed intake, prolactin, growth hormone and
     glucocorticoid response to cows to supplemental light.
     J. Dairy. Sci., 64, 1571-1678.

Revault, J.P., Daveau, A. and Ortavant, R. 1981. Evidence
     for a sensitive phase for prolactin secretion in rams.
     In: "Photoperiodism and reproduction". (Ed. R.
     Ortavant, J. Pelletier and J.P. Ravault) (INRA Publ).
     pp 135-146.

Schanbacher, B.D. and Grouse J.D. 1981. Photoperiodic regu-
     lation of growth. A photosensitive phase during light
     dark cycle. Amer. J. Physiol., 241, E1.

Terqui, M., Garnier, D.H., de Reviers M.M., Huet, S. and
     Pelletier, J. 1980. Structure chronologique du
     dialogue entre l'hypophyse et les gonades chez les
     mammifères domestiques. In: "Rythmes et reproduction".
     (Ed. R. Ortavant and A. Reinberg) (Masson, Paris)
     pp 59-72.

Thibault, C., Courot, M., Martinet, L., Mauleon, P., du Mesnil
     du Buisson, F., Ortavant, R., Pelletier, J. and Signoret,
     J.P. 1966. Regulation of breeding season and oestrous
     cycle by light and external stimuli in some mammals.
     J. Anim. Sci., Suppl. 25, 119-141.

DISCUSSION ON DR. TERQUI'S PAPER

<u>Dr. Heitzman</u>:   Is there a relationship between the intensity of the light and the response?

<u>Dr. Terqui</u>:   The animals received the same amount of light as in natural conditions. But the intensity may be important. The second difference in the experimental compared with natural conditions is that an abrupt change in light occurs at dawn and dusk and there is no decrease at sunset and sunrise.

<u>Dr. Heitzman</u>:   Presumably there must be a receptor to accept the light and it must work at specific wavelengths.

<u>Dr. Terqui</u>:   In ram we have no information of a precise wavelength effect such as exists for ducks. The artificial light used in our experiments has the same spectrum as that of natural light.

<u>Dr. Roche</u>:   In general neither intensity nor wavelength is that important in cattle based on Dr. Tuckers data from Michigan State, once you get about 50 - 100 lux.

<u>Dr. Schanbacher</u>:   Dr. Tucker suggests that ruminants have a sensitivity to light at a level of maybe 20 lux or above. He also commented on the differential between light and dark. With birds an intensity of 2 lux is sufficient to maintain egg production or weight gains of broilers. There are probably some species differences with regard to light intensity but the question of light quality is a separate one. Wavelength does come into the picture. We are talking about the optic nerve as well as direct light coming through the cranium.

<u>Dr. Boland</u>:   Is there any seasonal effect on when you started your experiment, or have you repeated it throughout the year?

<u>Dr. Terqui</u>:   We have done the experiment beginning with both long days and short days with the same photosensitive effects.

<u>Dr. Schanbacher</u>:   I would like clarification of your point that long days stimulate LH, FSH and testosterone. We have

the opposite effect. I am wondering if the difference between your results and ours has to do with time of year the animals go on to the experiment, or whether it relates to something which has yet to be discovered.

Dr. Terqui: We have done the experiment for two seasons, beginning in Dec - Jan, and June, and on the two occasions we have seen similar effects. There are many differences between your experiments and ours such as breed of rams, light intensity, and the time of dawn. In the June experiment, all the groups receiving 7 + 1 hours except 16-17 hr. group have a transient rise in testes diameter but only the group 16-17 hr. have testes variation similar to the group with natural day light variations.

Dr. Roche: In relation to prolactin, temperature is important. I do not know how important it is in this case.

Dr. Terqui: Low and high temperature modify prolactin secretion. In our experiments the temperature was not kept constant, but all pens were submitted to similar temperature variations.

Dr. Barenton: I would like to make a comment on the intensity of light. Low intensity is effective. If you consider that light stimulates hormones between the 16th and 17th hour after dawn, in our conditions in France, this implies that by the time the 16th and 17th hour is reached, light of low intensity is found which is effective.

Dr. Roche: This would be in the summer.

Dr. Barenton: Yes.

EFFECT OF EXTENDED PHOTOPERIOD ON THE GROWTH OF SHEEP

B. R. Brinklow and J. M. Forbes
Department of Animal Physiology and Nutrition
University of Leeds
Leeds LS2 9JT, England

ABSTRACT

Growing lambs kept for 3-4 months under artificial long photoperiods (16L:8D) grow significantly faster than those under short photoperiod (8L:16D). Much of the extra weight is in gut fill but carcass weights are also improved, especially with ad libitum feeding. Skeleton long photoperiod (7L:10D:1L:6D) has similar effects to long photoperiod; the effects of both can be blocked by pinealectomy. Continuous light has no effect and attempts to stimulate growth under farm conditions by extending natural photoperiod to 20L:4D with artificial light have proved to be unsuccessful. Photoperiod affects the levels of prolactin, cortisol and testosterone in blood but their roles in the mode of action of daylength effects on growth are not fully understood.

INTRODUCTION

In a seasonally breeding animal such as the sheep, any given phase of growth will normally take place at the same time of year in most individuals and thus growth rate and development are seasonally interrelated. Under natural conditions there are likely to be seasonal differences in the availability and quality of food, which will affect growth, and to which the animal is likely to have evolved adaptations.

In sheep, both growth and food intake are highest in the summer and lowest in the winter, even when given access to equally high quality food the year round. In following the course of voluntary intake of concentrate feeds in sheep from four months to four years of age (Suffolk x Finn-Dorset) Blaxter and Gill (1979) found mean intake to be 13% higher in July than in January. The amplitude of these fluctuations was found to be highest in the less improved Soay breed than in the lowland Suffolk x Finn-Dorset crossbreed (Kay, 1979).

These effects have been shown to be due to changes
in photoperiod length by demonstrations that the annual
cycle of growth and food intake could be compressed into
six months using suitable photoperiodic manipulation
(Brown et al 1979; Kay, 1979; Kay and Suttie, 1980).
It was also found that castrated males (Kay, 1979) show
a lower overall level of intake and a smaller amplitude
of the appetite cycle than intact males. Also, in the
case of deer, hinds show a lower level of intake with a
smaller amplitude fluctuation than stags (Suttie, 1981).

LONG vs SHORT PHOTOPERIODS

The growth and carcasses of lambs exposed to fixed
photoperiods of either 16L:8D or 8L:16D, after an initial
two week period on 12L:12D, have been studied in detail
in a series of experiments at the University of Leeds.
The first of these involved 72 April-born female and
castrated male lambs sired by Oxford and Suffolk rams
out of Finn-Blackface ewes and lasted for 16 weeks starting
at six months of age (Forbes et al 1979a, Experiment 1).
Half were fed ad libitum on a concentrate ration while
the others were restricted to 70g/kg liveweight $^{0.73}$/day
(Designed to give a growth rate of approximately 100 g/day
(A.R.C. 1965). Live weight gains were significantly
higher in animals on long photoperiods at both levels of
feeding (ad libitum: 225 vs 151 g/day; restricted: 200 vs
121 g/day). Carcass weights were greater by at least
1.5 kg in favour of long photoperiods, but some of the
advantage in liveweight was in the extra gut fill of the
restricted fed lambs. Internal fat deposits were not
affected by photoperiod, but the carcasses tended to be
larger following exposure to long photoperiods. There
were also effects on the weight of pelt and head. There
was no effect of photoperiod on the composition of a
standard three-rib sample joint. These results suggest
that the difference in growth was in all major tissues.

The results of a similar experiment have been published by Schanbacher and Crouse (1980). Twelve intact
and 12 castrated male lambs aged 10 weeks offered concentrate food ad libitum were exposed to 16L:8D and the same
number to 8L:16D. Over 12 weeks the weight gains were
significantly greater in rams than in wethers (375 vs
323 g/day) and in long compared with short photoperiods
(378 vs 320 g/day) with no interaction between 'sex' and
photoperiod. Carcass weights were elevated following long
photoperiod treatment being on average 2.9 kg heavier than
under short photoperiods.

The results of a second experiment at Leeds with 24
Oxford- or Suffolk-sired crossbred lambs of a similar age
and at a similar time of year to experiment 1 and fed at
the restricted level, were not as clear cut as those of
the first experiment (Forbes et al, 1979a, experiment 2).
Those kept under 16L:8D had weight gains of 144 compared
with those under 8L:16D of 127 g/day. Some of this
difference was again accounted for by gut fill (Lambs on
16L:8D on average having 0.9 kg more gut contents ($P < 0.01$)
and carcasses were 1.1 kg heavier in favour of long
photoperiods though this was not statistically significant.
There was little significant effect of photoperiod on
carcass dimensions or composition though the sample joints
in long photoperiods tended to have a higher proportion
of muscle and a lower proportion of fat.

A third experiment with 24 castrated male animals
of Suffolk-sired crossbreed compared 16L:8D with 8L:16D
and ad libitum with restricted feeding (Forbes et al, 1981,
experiment 3). The experiment began when the animals were
approximately 5 months of age in late September and lasted
for 12 weeks. Once again the liveweight gains were
elevated in the long photoperiods with both ad libitum
(180 vs 142 g/day) and restricted (102 vs 90 g/day)
feeding, due mainly to a significant effect of photoperiod

in the third month. Gut fill was increased, especially
in the restricted fed animals (P < 0.05) and carcasses
were on average 1.0 kg heavier in long photoperiod animals
though this effect was again not significant; carcass
dimensions and composition were not significantly affected
by lighting regime. When 9-month-old female Suffolk-sired
crossbred lambs were slaughtered 4 and 7 weeks after
exposure to long and short photoperiods (Forbes et al, 1981,
experiment 4) the weight of gut contents was significantly
different at both times showing that the effect of photoperiod on gut fill occurs within one month.

Of the 3 experiments reviewed above which included
ad libitum fed animals, the first (Forbes et al, 1979a,
experiment 1) showed that at the end of 16 weeks of
treatment those under long photoperiods were eating 2.0
kg/day while those under short photoperiods ate 1.7 kg/day
(P < 0.01). In the second experiment from Leeds (Forbes
et al, 1981, experiment 3) the effect was smaller (1.32 vs
1.19 kg/day) and not statistically significant. The
experiment of Schanbacher and Crouse (1980) showed the
intake of concentrate feed to be 1.77 and 1.63 kg/day in
rams and 1.56 and 1.45 kg/day in wethers under 16L:8D
and 8L:16D respectively (P < 0.01).

Pair feeding experiments and the effects of pinealectomy

Because of the feeding regimes used in these experiments, faster growing animals were given more food, or in
the case of ad libitum feeding, ate more voluntarily. In
an attempt to eliminate these effects two experiments were
carried out using a pair feeding regime in which animals
were paired by liveweight, one of each pair being allocated
to each photoperiod group and from then on feeding carried
out at 70 g/kg liveweight$^{0.73}$/day using the mean liveweight of the pair (Forbes et al, 1981, experiment 1 and 2).

The first of these experiments was carried out at the same time of year and with similarly aged wether lambs of Suffolk and Oxford-sired crossbreeds as the previously reviewed experiments from Leeds. This showed a significant effect of photoperiod on liveweight gain in the second and third months (of 15 weeks) of treatment. In this experiment, 4 animals in each group were pinealectomised (Px) and 4 sham Px (Roche & Dziuk, 1969); although the overall differences in liveweight gain in the non-operated and sham Px animals were statistically significant, the effect was much smaller and not statistically significantly different in the Px animals. Analysis of variance showed an almost significant interaction between photoperiod and surgical treatment ($P = 0.064$). (This is in agreement with the results of a pilot experiment of Forbes (1975) who found, with a small number of animals, that Px removed the effect of photoperiod on growth in lambs). The over-all difference in liveweight gain was not significant (137 vs 121 g/day) ($P < 0.07$) for long vs short photo-periods respectively) and was largely due to gut fill (6.3 vs 5.2 kg, $P < 0.02$). Carcasses from the sheep overall under 16L:8D were 0.6 kg heavier though this was not a significant difference, and the carcasses of the Px animals were in fact 0.2 kg heavier in short photoperiod (SP) than in long photoperiod (LP) (Brown, 1979). In this experiment the lambs on LP had a significantly higher proportion of bone and muscle and less fat in the 3-rib joints and a significantly lower backfat thickness and weight of internal fat deposits. Pinealectomy had little discernible consistent effect on these parameters except internal fat where the differences were smaller in the Px animals. Largely similar results were obtained from the second of the pair feeding experiments in which 24 Suffolk-sired crossbred lambs were exposed to photoperiod treatment between 2.5 and 5 months of age (May to August).

The liveweight gains were significantly higher in the
second month in the LP lambs though this difference
was not significant overall (175 vs 162 for LP vs SP) and
in fact in the third month there was a significant
difference in the opposite direction. When initial liveweight was used as a covariate in the analysis of variance
the overall liveweight gain was significantly greater in
LP (P < 0.05). There was a significantly higher gut fill
in the LP animals with significantly less internal fat
deposition. There were no consistent differences in
carcass composition in this experiment. An overall
conclusion from these 2 experiments was that long photoperiods stimulated the growth of earlier developing
tissues (bone and muscle) and, because of the exactly
equal weights of food offered, those animals exposed to
LP were relatively under-fed and had less energy available
for fat deposition after the increased requirements for
non-fat growth had been met.

SKELETON LONG PHOTOPERIODS

The photoperiodic time measurement involved in these
photoperiodic effects on growth has been shown to be by
a coincidence mechanism (Bunning, 1964; Pittendrigh, 1976)
where the effect of a long photoperiod is due to light
occurring in a photosensitive phase of a circadian rhythm
of light sensitivity. This was confirmed by experiments
using skelton long photoperiods' (SLP) in which a
relatively short daily duration of light is split to
encompass the period of a long photoperiod. Schanbacher
and Crouse (1981) exposed ram lambs to either 8L:16D
16L:8D or 7L:9D:1L:7D from 2.5 months of age for 11 weeks,
with a concentrate feed offered ad libitum. The results
of the experiment can be seen in table 1 and demonstrate
that the SLP was equivalent to the LP in giving higher
weight gains, carcass weights and food intakes and better

food conversion efficiency. There was no effect on
carcass quality.

TABLE 1. Means of traits of ram lambs after exposure to
different photoperiods for 11 week.

Photoperiod	Average daily liveweight gain (g)	Feed intake/ animal (kg)	Feed conversion efficiency kg feed/kg lwtgain	Carcass wt (kg)
8L:16D*	345*	125*	4.6*	21.9*
16L:8D	417∅	141∅	4.1∅	25.1∅
7L:10D:1L:6D	442∅	138∅	4.2∅	24.4∅

*∅ Means without a common symbol differ significantly
(P < 0.01). Data from Schanbacher and Crouse (1981).

We have also compared the effects on growth of lambs
of 7L:10D:1L:6D against 8L:16D in 4 experiments of the same
general type as those described above using Suffolk-sired
crossbred lambs. In all cases a concentrate feed was offered
at 70 g/kg LW $^{0.73}$/day (Brinklow, Forbes & Jones, 1982).
The first experiment involved female and castrated male
lambs from approximately 3 to 6 months of age beginning
in May. Liveweight gains were 105.0 and 126.6 g/day with
FCE values of 8.3 and 6.8 kg food/kg liveweight
gain for SP and SLP respectively (both P < 0.05) with
no significant effects on carcass size (though there was
a tendency for SLP animals to be larger) or sample joint
composition. The second experiment was similar in design
to the first though the animals were 6 months old at the
beginning of the experiment in September. The weight gains
were again higher in the SLP animals (78.9 vs 95.2 g/day,
P < 0.07). These gains were much lower than in the previous
experiment and were accompanied by higher food conversion
values (15.8 and 13.2 kg food/kg live weight gain, P < 0.05)
even though the level of feeding and source of food were
the same. This may indicate relative undernutrition due

to poor digestibility of the ration (possibly due to
increased rate of passage due to fine grinding of
constituents) and to this experiment the 3-rib sample
joins were significantly leaner (P < 0.01 for the differences in the proportion of muscle and of fat between the
2 photoperiods) in the SLP group, a result which may be
consistent with earlier findings in the pair feeding
experiments.

Effects of Pinealectomy

The second pair of experiments in this series looked
also at the effects of Px. The first of these used female
and castrated male lambs and began in February with the
lambs at 11 months of age. There was no significant effect
of photoperiod in either the Px or non-Px animals over the
next 8 weeks on weight gain or on carcass dimensions or
size at the end of this time. The final experiment used
intact and castrated male lambs of a similar age and at a
similar time of year to the first experiment in the series.
Though not significantly different the weight gains were
larger in the SLP than in the SP with the differences
being much smaller in the Px than the non-Px animals.
Taking the 4 experiments overall and removing the effects
of sex and experiment by analysis of variance, the effect
of photoperiod on mean daily liveweight gain was highly
significant (P < 0.005), SLP > SP in the non-Px animals,
and not significantly different in the Px animals. There
was, however, no overall tendency for an improvement in
carcass weight in SLP and the difference between this and
the increased liveweight at slaughter was not accounted
for by increased gut fill. In respect of the carcass
weights these results differ from those of Schanbacher and
Crouse (1981) where there was a significant increase in
carcass weight though these animals were fed ad libitum
and the weight gains were also very much larger.

CONSTANT ILLUMINATION

The effects of constant illumination throughout the

night during the winter (natural photoperiods of 9-12 h light/day) was found by Hackett and Hillers (1979) to have no effect on growth rate in sheep. Hoersch et al (1961) found that constant illumination gave similar growth rates to those found under 12L:12D and Moose and Ross (1962) found that liveweight gain in lambs was slower in constant illumination than in natural lighting, though the time of year was not mentioned. Similar results have been found by Peters et al (1980) and Leining et al (1979) for both growth and the secretion of PRL in cattle and by Brinklow (1982) for PRL in sheep.

PRACTICAL APPLICATION

Practical application of this effect of photoperiod on growth has been studied by Jones et al (1982). Four trials were carried out with groups of 6-month-old lambs kept in barns open on at least one side, in which natural light was supplemented at 50 lux intensity from dusk until 22.00 h and from 02.00 h until dawn between September and January. There were no significant effects of 20L:4D on the weight of saleable carcass in any trial though there was a large effect on live weight gain in one trial which gave slightly heavier carcasses, higher carcass grades and a significantly higher financial return. In another trial the lambs with supplemental lighting were leaner. Both of these trials used Suffolk-sired crossbred lambs, the other 2 trials in which there were no significant effects on weight gain or carcass weight were of the Dorset Down x Welsh crossbred. Overall the investment in equipment and electricity was not warranted by the slightly increased returns.

One important consideration that is relevant to the interpretation of the results from these trials and to other experiments (such as those of Hackett and Hillers (1979)) where natural light is supplemented with artifical light is intensity. It has been demonstrated that, for

the control of melatonin secretion in rodents, a given
intensity of light may be seen as either 'light' or 'darkness' depending on the relative intensity of the alter-
nating photoperiod (Lynch et al, 1981). Artificial
lighting does not normally approach the intensity of
natural light and if this phenomenon were relevant to
other photoperiodic mechanisms, or if in fact melatonin
is controlled in the same way and is involved in photoperiodic mechanisms in sheep, then it may be difficult to
determine whether the supplemental lighting was seen as
either day or night by the animal. It may thus be more
effective (and also less expensive) to extend winter
photoperiods in the manner of a skeleton long photoperiod
and supply a pulse of light during the night, i.e. removing
the natural-light/artificial-light interface. The only
trials of this type published were by Hackett and Hillers
(1979) who found, however, that lambs subjected to
natural photoperiods in Washington State, U.S.A. (latitude
approximately 52°N) supplemented with 2 hours of light
between 23.00 and 07.00 h during April to June, July to
September or September to January showed no significant
differences on growth compared to those animals with no
supplement, at any of these times of the year.

POSSIBLE MODES OF ACTION
     The pineal gland is known to be involved in many
photoperiodic effects including the secretion of the
reproductive hormones and prolactin in sheep (see Lincoln
and Short, 1980) and there is evidence that it has its
effect at least partially by its secretion of melatonin
during periods of darkness. The effect of Px on growth
in lambs has been studied at Leeds both in experiments
using long and short, and skeleton long and short
photoperiods.
     In all of these experiments where there was an effect
of photoperiod on growth in non-Px animals this was
reduced or abolished in Px animals.

Of the hormones associated with growth ciruclating prolactin (PRL) has been shown to be closely related to photoperiod length in sheep with those photoperiods that cause increased growth also causing increased PRL and _vice_ _versa_. However, growth was not affected in all cases when PRL showed a response (e.g. Brinklow, Forbes & Jones, experiment 3). It seems that photoperiod may affect growth predominantly in the early rapid phase of growth (the animals in the above-mentioned experiment being approximately one year old) whereas prolactin is affected by photoperiod at all ages. Also in cattle Perers et al (1980) have shown an effect of increased growth in artificially extended winter photoperiods in the absence of any effect on PRL due to low temperature. Both Brown et al (1976) and Ravault et al (1977) showed no effect of the PRL secretion blocker 2-bromo- -ergocryptine (CB154) on the growth of lambs in long photoperiods though there were no short photoperiods groups for comparison. Eisemann et al (1982) did, however, find an effect of CB154 on growth of weanling lambs in (16L:8D) but no effect of prolactin supplementation in (8L:16D) and in fact, no significant difference between control animals on either photoperiod, though the experiment was a relatively short duration (9 weeks). Ohlson et al (1981) found that autoimmunisation against PRL reduced weight gains in yearling rams in the summer, and Brinklow and Forbes (1982) found that continuous intravenous infusion of ovine PRL for 10 days into four 6-month-old castrated male, Suffolk-sired crossbred lambs maintained in continuous darkness (producing low endogenous PRL levels) significantly increased nitrogen retention.

Of the other hormones that have been implicated in the control of growth, insulin, growth hormone and thyroid hormones have been shown to be not directly affected by photoperiod in sheep (Forbes et al, 1979b).

The glucocorticoid cortisol has been found to be

271

decreased in skeleton long photoperiods when there was
an effect of photoperiod on growth (Brinklow, Forbes and
Jones, experiments 1 and 4) and not when there was no
effect on growth (experiment 3). This hormone has also
recently been implicated in the mode of action in sheep
of the anabolic agent trenbolone acetate (Thomas and Rodway,
1982).

The effects of testosterone on the growth of sheep
are well established with rams growing faster than wethers
(Everitt and Juny, 1966a,b; Schanbacher et al, 1980) and
testosterone therapy being capable of restoring the growth
of wethers to the level of that of rams. The effect of
photoperiod on testosterone tends to be opposite to that
on growth, however, with increased levels being found in
the autumn and winter under the stimulation of short
photoperiods.

CONCLUSIONS

It is clear that there are marked effects of both
artificial long photoperiods and skeleton long photoperiods
on the growth of lambs, which are shown most clearly with
ad libitum feeding. Attempts to harness these observations
to the practical situation have so far yielded disappointing
results.

Prolactin levels are elevated in most situations in
which growth is stimulated but a causal relationship is
difficult to establish. Cortisol is protein catabolic
and is depressed in skeleton long photoperiods, though
testosterone which is anabolic is also depressed in long
photoperiods. It is possible that the pineal gland is
involved in these responses but the pathway from eyes
to growing tissues is still far from clear.

ACKNOWLEDGEMENTS

We thank Mr. R. Jones for invaluable assistance in
the planning and execution of the experiments at Leeds
University. Much of the work summarised here was

supported by grants from the Agricultural Research
Council, to whom we are most grateful.

REFERENCES
A.R.C. 1965. The Nutritional Requirements of Farm Livestock.
    No. 2, Ruminants. Agricultural Research Council,
    London.
Blaxter, K.L. and Gill, J.C. 1979. Voluntary intake of
    feed and equilibrium body weight in sheep. Proc.
    Nutr. Soc., $\underline{38}$, 150A.
Brinklow, B.R. and Forbes, J.M. 1982. Anim. Prod. (full
    paper in preparation).
Brinklow, B.R., Forbes, J.M. and Jones, R. 1982. Prolactin
    infusion causes increased nitrogen retention in
    lambs in continuous darkness. Br. J. Nutr. (In press).
Brown, W.B. 1979. Daylength effects on the growth of lambs:
    involvement of hormones and the pineal gland. Ph.D.
    thesis, University of Leeds.
Brown, W. B., Forbes, J.M., Goodall, E.D., Kay, R. N.B.,
    and Simpson, A.M. 1979. Effects of photoperiod on
    food intake, sexual condition and hormone concen-
    trations in stags and rams. J. Physiol. $\underline{296}$, 58P-59P.
Brown, W. B., Driver, P. M., Jones, R., and Forbes, J. M.
    1976. Growth, prolactin and growth hormone in lambs
    treated with CB154. Proc. Soc. Endocr. 69, 47P.
Bunning, E. 1964. The Physiological Clock. Springer-
    Verlag, Berlin.
Eisemann, J. H., Bauman, D.E., Hogue, D.E. and Travis, H.F.
    1982/3. J. Anim. Sci. In press.
Everitt, G.C. and Jury, K. E. 1966a. Effects of sex and
    gonadectomy on the growth and development of
    Southdown x Romney cross lambs. Part 1. Effects on
    liveweight growth and components of liveweight.
    J. Agric. Sci. $\underline{66}$, 1-14.
Everitt, G.C. and Jury, K.E. 1966b. Effects of sex and
    gonadectomy on the growth and development of Southdown
    x Romney cross lambs. Part II. Effects on carcass
    grades, measurements and chemical composition. J. Agric.
    Sci. $\underline{66}$, 15-27.
Forbes, J. M., El Shahat, A. A., Jones, R., Duncan, J.G.S.
    and Boaz, T.G. 1979a. The effects of daylength on
    the growth of lambs. 1. Comparisons of sex, level
    of feeding, shearing and breed of sire. Anim. Prod.,
    $\underline{28}$, 33-42.
Forbes, J.M., Brown, W.B., Al-Banna, A.G.M. and Jones, R.
    1981. The effect of daylength on the growth of lambs.
    3. Level of feeding, age of lamb and speed of gut-fill
    response. Anim. Prod. $\underline{32}$, 23-28.
Hackett, M.R. and Hillers, J.K. 1979. Effects of artificial
    lighting on feeder lamb performance. J. Anim. Sci.,
    $\underline{49}$, 1-4.

Hoersch, T. M., Reineke, E. P. and Henneman, H. A. 1961.
    Effect of artificial light and ambient temperature
    on the thyroid secretion rate and other metabolic
    measures in sheep. J. Anim. Sci., 20, 358-362.
Jones, R., Forbes, J. M., C.F.R. Slade and Appleton, M.
    1982. The effect of daylength on the growth of
    lambs. 4. Daylength extension to 20 h under practical
    conditions. Anim. Prod. 35, 9-14.
Kay, R.N. B. 1979. Seasonal changes of appetite in deer
    and sheep. A.R.C. Res. Rev. 5, 13-15.
Kay, R.N.B. and Suttie, J.M., 1980. Relationship of
    seasonal cycles of food intake and sexual activity
    in Soay rams. J. Physiol. 310, 34-35P.
Leining, K.B., Tucker, H.A. and Kesner, J.F. 1980.
    Growth hormone, glucocorticoid and thyroxine response
    to duration, intensity and wavelength of light in
    prepubertal bulls. J. Anim. Sci., 51, 932-942.
Lincoln, G.A. and Short, R.V. 1980. Seasonal breeding :
    Nature's contraceptive. Rec. Prog. Horm. Res. 36,
    1-52.
Lynch, H.J., Rivest, R.W., Ronsheim, P.W. and Wurtman, R.J.
    1981. Light intensity and the control of melatonin
    secretion in rats. Neuroendocr. 33, 181-185.
Moose, M.G. and Ross, C.V. 1962. Response of fattening
    lambs to artificial light. J. Anim. Sci. 21, 1040.
Ohlson, D. L., Spicer, L.J. and Davis, S.L. 1981. Use
    of active immunisation against prolactin to study
    the influence of prolactin on growth and reproduction
    in the ram. J. Anim. Sci. 52, 1350-1359.
Peters, R.R., Chapin, L.T., Emery, R.S. and Tucker, H.A.
    1980. Growth and hormonal response of heifers to
    various photoperiods. J. Anim. Sci. 51, 1148-1153.
Pittendrigh, C.S. 1972. Circadian surfaces and the diversity
    of possible roles of circadian organisation in
    photoperiodic induction. Proc. Nat. Acad. Sci.
    U.S.A. 69, 2734-2737.
Revault, J.P., Courot, M., Garnier, D., Pelletier, J.,
    Terqui, M. 1977. Biol. Reprod. 1, 192-197.
Roche, J.F. and Dziuk, P.J. 1969. A technique for
    pinealectomy of the ewe. Am. J. Vet. Res. 30,
    2031-2035.
Schanbacher, B.D. and Crouse, J.D. 1980. Growth and per-
    formance of growing-finishing lambs exposed to long
    or short photoperiods. J. Anim. Sci. 51, 943-948.
Schanbacher, B.D. and Crouse, J.D. 1981. Photoperiodic
    regulation of growth; a photosensitive phase during
    light-dark cycle. Am. J. Physiol. 241, E1-E5.
Schanbacher, B.D., Crouse, J.D. and Ferrell, C.L. 1980.
    Testosterone influences on growth performance carcass
    characteristics and composition of young market lambs.
    J. Anim. Sci. 51, 685-691.
Suttie, J.M. 1981. Ph.D. thesis, University of Aberdeen.
Thomas,K.M. and Rodway, R.G. 1982. Proc. Nutr. Soc.
    (In press).

DISCUSSION ON DR. FORBES'S PAPER

Professor Karg: Why do you associate prolactin with gut fill? Do you have any speculation on the thyroid gland? Thyrostatics increase gut fill and there are also seasonal variations in thyroid function.

Dr. Forbes: We have only measured T4 and it would be better to measure T3 as well, but we could find no significant effect of long versus short days on T4. So I would not like to exclude your suggestion but what bit of evidence we have would not help us.

Dr. Galbraith: How did you decide on the energy level of the diet for feeding the animals? It strikes me that your animals were gaining 100g per day which is not much above the maintenance requirement. You have probably not given yourself as much opportunity of altering the rate of gain by giving them such a ration. If you fed them to gain 300g per day you would have had much more scope of picking up possible treatment effects.

Dr. Forbes: Yes, but it is almost certain we would have run into problems with some animals not eating all their ration. We wanted to avoid this because we did not want any differences in feeding.

Dr. Galbraith: A greater level of energy intake could be obtained, while remaining within the limits of appetite for dry matter intake, by increasing the metabolisable energy content of the ration. This may be achieved by, decreasing the roughage and increasing the concentrate components.

Dr. Boland: Lambs on 16hr light were growing faster after 5 - 6 weeks. What happens if you switch them back to 8hr light at that stage?

Dr. Forbes: The animals "know" that they are in different days within a couple of days. Prolactin changes rather rapidly.

HORMONAL AND PHOTOPERIODIC CONTROL OF GROWTH

B. D. Schanbacher

Roman L. Hruska U.S. Meat Animal Research Center
U.S. Department of Agriculture
Box 166
Clay Center, Nebraska 68933
USA

ABSTRACT

The growth and performance of growing-finishing steers receiving growth stimulants via commercially-available subdermal implants are discussed in relation to the endogenous steroids secreted by the bovine testis. Although testosterone is quantitatively the most important steroid secreted by the mammalian testis and is known to possess potent anabolic properties, its androgenic and oestrogenic metabolites may mediate its anabolic effects in vivo. The potency by which androgen-oestrogen combination implants (e.g., Forplix and Revalor) increase weight gain in steers lends credence to this hypothesis.

The effects of contrasting photoperiods on weight gain of young lambs fed ad libitum have been studied. In summary, a 16-h day (long photoperiod) enhances the growth rates of ram lambs, ewe lambs and wether lambs when compared to lambs exposed to an 8-h day (short photoperiod). The stimulatory effects of 16-h of continuous light can also be achieved in lambs exposed to 8-h of light per day if one-hour of light is provided as a light flash at hour 17 after dawn. Photoperiodic manipulation of lamb growth allows producers to maximize the utility of a confinement facility for finishing lambs.

INTRODUCTION

Increased production efficiency is essential for the livestock producer if animal products (e.g. meat, milk and eggs) are to compete successfully as a basic food group in the human diet (Harper et al, 1980). Domestic animals which provide most of the animal protein for the human diet are managed under increasingly intensified systems. Coupled with these intensified systems of animal production are problems which challenge even the most progressive livestock producer. For example, sexual and aggressive behavior in confined, densely-populated groups of animals result in decreased production efficiency. Castration of the intact male has been used effectively to remove the negative influence of male sexual behavior on animal performance but this management practice results in decreased production efficiency because of the removal of the naturally-occurring anabolic steroids secreted by the testis (Seideman et al, 1982).

In addition to the problems associated with intensified systems of animal production are the advantages of feed storage and handling, reduced labor requirements and the possibilities of increased production efficiency through environmental manipulation (Tucker and Ringer, 1982). The first part of this paper discusses the consequences of castrating meat producing animals and outlines some of the guidelines used in formulating implants for effective steroid replacement therapy. The second part of this paper presents evidence that manipulation of daylength through exposure to artificial photoperiod can have dramatic effects on the efficiency of lamb production.

STEROID CONTROL OF GROWTH

The Testes (Testosterone?)

The more rapid growth, more efficient feed utilization and leaner carcasses of intact males when compared to castrated males are assumed to be attributable to testicular steroid hormones (Field, 1971; Seideman et al, 1982), in particular, testosterone (Schanbacher and Ford, 1976; Schanbacher et al, 1980). The probability that testosterone is responsible for the growth differential between intact and castrate ram lambs is supported by the experimental data shown in Table 1. Immuno-neutralization of endogenous testosterone in ram lambs by active immunization against testosterone resulted in rates of gain that were comparable to surgically-castrated lambs. Note that control immunization against bovine serum albumin (BSA) was not detrimental to lamb growth. In contrast, castrate lambs (wethers) given testosterone replacement therapy

TABLE 1   Relative growth rates of ram lambs immunized against testosterone and wether lambs implanted with testosterone-filled capsules.

Treatment	n	ADG (gm/day)
Ram	12	414[a]
Ram (anti BSA)	12	420[a]
Ram (anti T)	12	343[b]
Wether	12	355[b]
Wether (+T)	12	402[a]

[ab] Means differ significantly (P<0.01). From Schanbacher et al, 1980 and Schanbacher, 1982.

by subdermal testosterone capsules had rates of gain that were significantly greater than nonimplanted castrates and comparable to intact ram lambs. These data provide strong support for the suggestion that testosterone is the principal anabolic steroid secreted by the testis.

Although the predominant steroid secreted by the mammalian testis, testosterone is known to be metabolized in peripheral and central tissues to the more biologically potent steroids, 5α-dihydrostestosterone (DHT) and oestradiol-17β (E$_2$). The enzymes responsible for these metabolic conversions, 5α-reductase and the aromatase complex (Figure 1), are found in extragonadal tissues (Callard et al, 1978) along with receptors whose specificities are purely androgenic (Krieg, 1976) or oestrogenic (Dionne et al, 1980). These findings suggest that the anabolic effects of testosterone may be elicited by one or more of its metabolites. While testosterone remains one of the most powerful anabolic agents known, implants with primarily androgenic or oestrogenic activities are commercially prepared and sold for use in the livestock industry.

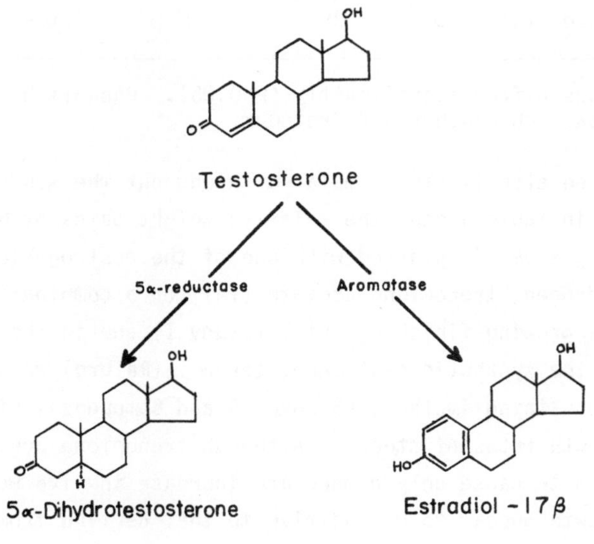

Fig. 1   Metabolism of testosterone to 5α-dihydrotestosterone (DHT) and oestradiol-17β (E$_2$).

278

Implants (Exogenous steroid?)

Numerous investigations have shown that synthetic oestrogen implants increase fat deposition in bulls and decrease fat deposition in steers (Bailey, 1966; Harte, 1969). While oestrogens stimulate weight gains in steers, their effects on weight gain in bulls are negligible (Baker and Arthaud, 1972). The data in table 2 show our recent findings regarding the use of commercially-available implants containing oestrogenic compounds in growing-finishing steers adapted to high grain diets in the U.S.A. Ralgro and Synovex-S implants were readministered twice so that

TABLE 2   Pasture-feedlot gains for implanted steers.

Treatment	n	205-day ADG (kg/day
Control	37	1.07[a]
Ralgro (3X)	38	1.22[b]
Synovex-S (3X)	39	1.16[c]
Compudose (1X)	38	1.16[c]

[abc] Means differ significantly (P<0.05). Unpublished data, Schanbacher and Brethour.

the steers would be effectively stimulated throughout the study (i.e. 205 days). The data in table 3 show the relative weight gains at 68 and 109 days for finishing steers implanted with one of the oestrogenic implants, the synthetic androgen, trenbolone acetate (TA), or a combination of the two. In both the growing-finishing trial (Study 1) and in the finishing trial (Study 2), the synthetic oestrogen, zeranol (Ralgro) as well as the two oestradiol-containing implants (Synovex-S and Compudose) significantly increased the growth rates of steers. Although trenbolone acetate (Finaplix) appears to cause only a moderate increase in live weight gain, its effect on growth appear to be additive to that derived from the oestrogenic implants. The apparent additive effect of combined androgen-oestrogen treatment (i.e., Forplix (TA + zeranol) and Revalor (TA + E_2)) supports the hypothesis of separate anabolic effects of androgen and oestrogen.

TABLE 3   Feedlot gain for implanted steers.

Treatment	n	68-day ADG (kg/day)	109-day ADG (kg/day)
Control	17	0.96[a]	1.23[a]
Finaplix	18	1.07[b]	1.25[a]
Ralgro	17	1.08[b]	1.27[ab]
Synovex-S	18	1.12[bc]	1.32[b]
Compudose	18	1.13[bc]	1.32[b]
Forplix	18	1.16[c]	1.34[bc]
Revalor	18	1.20[c]	1.37[c]

[abc] Means differ significantly (P<0.05).
Unpublished data, Schanbacher and Brethour.

Biosensitivity (Exogenous steroid?)

The need for a more potent bioassay to determine the suitability of
a newly-formulated implant or to monitor the legal and illegal uses of
exogenous steroids in livestock production has caused us to evaluate the
feedback sensitivity of pituitary luteinizing hormone (LH) secretion to
different anabolic steroids. In the castrate ram, different components
of the LH secretory system are affected as testosterone is administered
in increasing dosages (D'Occhio et al, 1982). From these data and those
of a similar study with steers (Schanbacher et al, 1983), it has been
suggested that analyses of LH pulse frequency and amplitude may be a
useful tool for evaluating the biopotency of the anabolic steroids.
Representative LH secretory profiles from a control steer (S) and steers
implanted with Compudose (E_2), Finaplix (TA) or given 5α-dihydro-
testosterone via intramuscular injection (DHT) are shown in figure 2.
The mean serum LH concentrations in response to a 1 μg intravenous
challenge dose of luteinizing hormone releasing hormone (LHRH) are shown
in table 4 for these treatment groups. Combined, it appears that mean
serum LH and the LH response to exogenous LHRH may provide a dependable
and sensitive assay for monitoring anabolic steroid use in the castrate
male. In this regard, it is interesting to note that commercially-

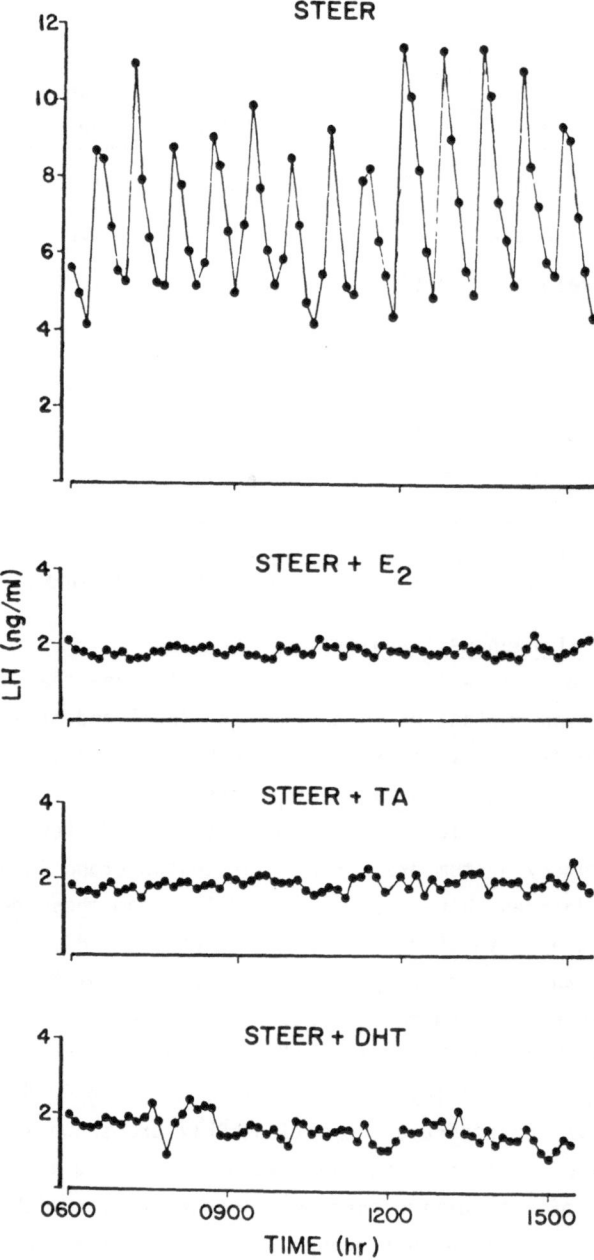

Fig. 2   Representative LH secretory profiles of steers, steers implanted with Compudose (E₂), steers implanted with Finaplix (TA) and steers administered dihydrotestosterone (DHT) by im injection.

TABLE 4   Mean serum LH concentrations and LH response to a
1 μg i.v. challenge dose of LHRH in nonimplanted steers and
steers treated with E_2 (Compudose), TA (Finaplix) and DHT.

Treatment	n	Mean LH (ng/ml)	LH respnse to LHRH (ng/ml)
Steers	5	7.8 ± 0.6^a	13.2 ± 1.5^a
Steers +E_2	5	2.1 ± 0.6^b	10.6 ± 1.5^a
Steers +TA	5	2.5 ± 0.6^b	3.0 ± 1.5^b
Steers +DHT	5	2.2 ± 0.6^b	1.6 ± 1.5^b

[a,b] Means ($\pm$SEM) differ significantly (P<0.01).
Unpublished data, Gettys and Schanbacher.

available implants are provided as growth stimulants at dosages which
suppress LH secretion in castrates.

PHOTOPERIODIC CONTROL OF GROWTH

Because animals are particularly susceptible to changes in their
environment, man has provided them with shelter. Manipulation of the
environment for several domestic species of livestock has increased their
rate of survival and their level of productivity. Although several
environmental factors are believed to affect animal performance, the
results of recent research suggest that photoperiod may be a major
regulator of efficient food production (Forbes, 1982; Schanbacher et al,
1982a; Tucker and Ringer, 1982).

The effect of contrasting long and short days on weight gain in young
lambs is illustrated in figure 3. Exposure of ram lambs, ewe lambs and
wether lambs to long days (16L:8D photoperiod) increased their growth
rates by approximately 21%, 12% and 15% respectively over that of lambs
exposed to short days (8L:16D photoperiod). Thus, long days appear to
stimulate weight gain in lambs independent of gonadal steroids or sex
condition (Schanbacher and Crouse, 1980; Schanbacher et al, 1982b). The
photoperiod effect also seems to behave independently of temperature. Ewe
lambs grew faster when exposed to long days regardless of whether ambient
temperatures were 8°C, 18°C or 32°C (Schanbacher et al, 1982b). The
ability to enhance growth rates in young market lambs by photoperiodic

PHOTOPERIOD EFFECTS ON WEIGHT GAIN

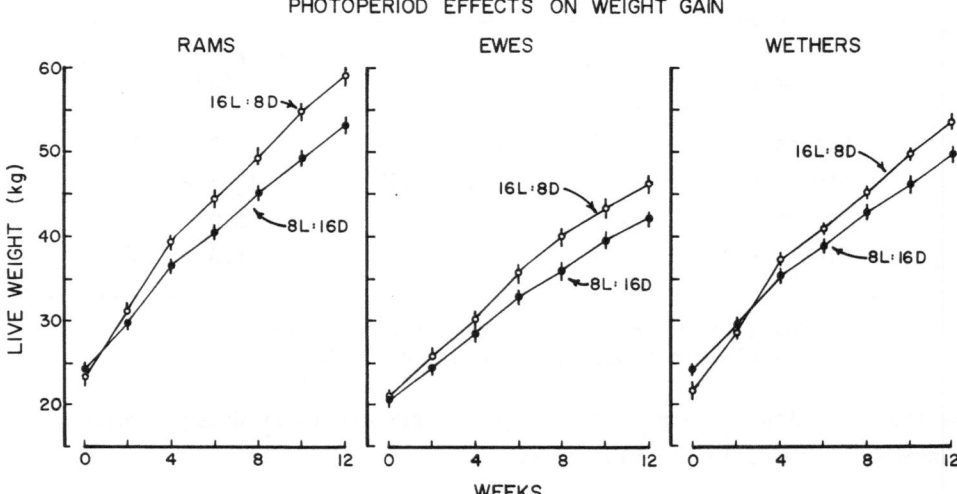

Fig. 3   Effects of contrasting long (16L:8D) and short (8L:16D) days
on weight gain of ram, ewe and wether lambs.

PHOTOSENSITIVITY IN LAMBS

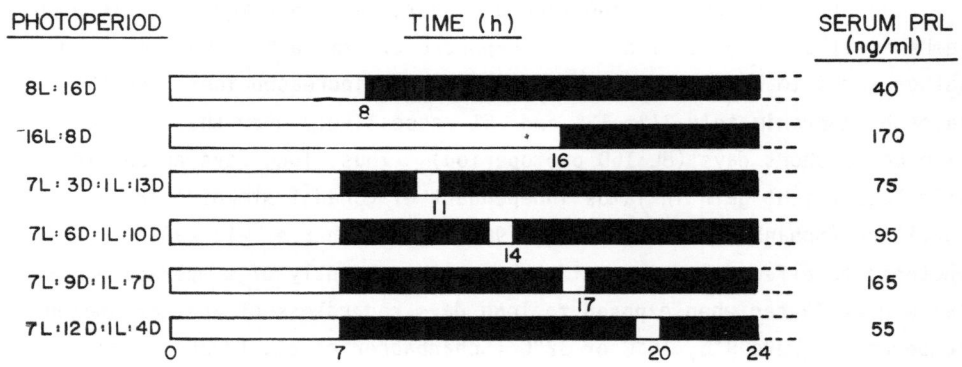

Fig. 4   Determination of a photosensitive phase during the light-dark
cycle in lambs via serum prolactin (PRL) concentrations.

manipulation is especially important to those producers who are attempting to maximize the utility of their intensively-managed confinement operation for fat lamb production. Furthermore, exposure of winter lambs in the northern hemisphere to artificial long days allows producers to market heavier lamb carcasses for the Easter trade.

A means to enhance growth rate in lambs without the overhead expenses of 16 hours of continuous artificial lighting have been successfully developed by extending the findings of Ravault and Ortavant (1977), who provided evidence for a photosensitive phase during the light-dark cycle of the sheep. Their work was extended in our laboratory to young lambs and prolactin data collected to verify the existence of a photosensitive phase between hour 16 and 17 of the light-dark cycle (figure 4). Based on our preliminary findings, 72 ram lambs were weaned at eight weeks of age and subsequently exposed to one of three treatment groups: a 8L:16D photoperiod (short day), a 16L:8D photoperiod (long day), or a 7L:9D:1L:7D photoperiod (hour-17 light flash). The lambs were exposed to the light treatments for approximately 11 weeks when they were slaughtered at a live mean weight of 50 kg. The results of this study (table 5) showed that lambs exposed to the 16L:8D or 7L:9D:1L:7D photoperiod out perform lambs exposed to the 8L:16D photoperiod. Interestingly, the lambs which received eight hours of light in the split photoperiod (light flash) group ate more feed and grew more rapidly than lambs exposed to the continuous eight hours of light. In fact, serum prolactin concentrations and

TABLE 5   Growth and performance data for ram lambs exposed to short (8L:16D), long (16L:8D) and split (7L:9D:1L:7D) photoperiods.

Treatment	Feed intake (kg/day)	Feed/ gain (kg/day)	ADG (gm/day)	Final weight (kg)	Carcass weight (kg)	Serum prolactin (ng/ml)
8L:16D	1.62[a]	4.6[a]	345[a]	45.4[a]	21.9[a]	123[a]
16L:8D	1.83[b]	4.1[b]	417[b]	51.0[b]	25.1[b]	302[b]
7L:9D:1L:7D	1.79[b]	4.2[b]	442[b]	53.0[b]	24.4[b]	328[b]

[ab] Means differ significantly (P<0.01).
From Schanbacher and Crouse, 1981.

performance traits for these lambs were similar to those of lambs exposed to the stimulatory 16L:8D photoperiod.

In a follow-up study, we have investigated the importance of the duration of the hour-17 light flash by monitoring 24-h serum prolactin responses to two light flashes of different duration and contrasting these responses to that of the regular 8L:16D and 16L:8D photoperiods (Table 6). The 24-h mean serum prolactin concentrations were found to be significantly elevated in all but the 8L:16D (short) photoperiod; implying that photo-stimulation not only occurs with the standard 16L:8D (long) photoperiod but also in photoperiods consisting of simple light flashes. Although both prolactin and testosterone are photoperiod responsive hormones and both are considered to be important anabolic agents of endogenous origin, it would appear from these data (table 6) that testosterone does not participate in the mechanism whereby light enhances growth rates in sheep. Although it might be similarly implied that prolactin is the mediator of photoperiod-induced acceleration of growth, a cause and effect relationship can not be determined from these results. Since neither pharmacological suppression of endogenous prolactin nor administration of physiological dosages of exogenous prolactin alters growth rate of young lambs (Eisemann, 1982) we conclude that photoperiod

TABLE 6   Hormonal responses of ram lambs exposed to four contrasting photoperiods.

Photoperiod	n	PRL (ng/ml)[a]	T (ng/ml)[a]
8L:16D	6	55 + 14[b]	3.7 + 0.3[b]
16L:8D	6	348 + 70[c]	2.0 + 0.2[c]
7L:9D:1L:7D	6	296 + 47[c]	1.6 + 0.2[c]
7L:9D:1'L:7.59D	6	267 + 55[c]	1.8 + 0.1[c]

[a] 24-h Means (+SEM) for serum prolactin (PRL) and testosterone (T).

[bc] Means differ significantly (P<0.01).
    Unpublished data, Schanbacher.

merely affects prolactin secretion, feed intake and growth rate in young lambs through a common mechanism(s) within the CNS. The influence of various lighting schemes on hormone secretory patterns in sheep and the mechanism(s) whereby light might affect growth rate in this species are subjects discussed more thoroughly at this workshop by Drs. Terqui and Forbes.

In conclusion, artificial lighting can be manipulated to improve the performance of confinement-reared lambs. This method allows for increased production efficiency which can 1) reduce feed costs, 2) defray overhead expenses associated with a confinement facility, and 3) permit more flexibility with regard to marketing.

REFERENCES

Bailey, C.M., Probert, C.L., Richardson, P., Bohman, V.R. and Chancerelle, J. 1966. Quality factors of the longissimus dorsi of young bulls and steers. J. Anim. Sci., 25, 504-508.

Baker, F.H. and Arthaud, V.H. 1972. Use of hormones or hormone active agents in production of slaughter bulls. J. Anim. Sci., 35, 752-754.

Dionne, F.T., Lesage, R.L., Dube, J.Y. and Tremblay, R.R. 1980. Estrogen binding proteins in rat skeletal and perineal muscles: In Vitro and In Vivo studies. J. Steroid Biochem., 11, 1073-1080.

D'Occhio, M.J., Schanbacher, B.D. and Kinder, J.E. 1982. Relationship between serum testosterone concentration and patterns of luteinizing hormone secretion in male sheep. Endocrinology, 110, 1547-1554.

Eisemann, J.H. 1982. Examination of the role of prolactin in growth and the photoperiod response. Ph.D. Thesis. Cornell University, Ithaca, New York.

Field, R.A. 1971. Effect of castration on meat quality and quantity. J. Anim. Sci., 32, 849-858.

Forbes, J.M. 1982. Effects of lighting pattern on growth, lactation and food intake of sheep, cattle and deer. Livestock Production Sci., 9, 361-374.

Krieg, M. 1976. Characterization of the androgen receptor in the skeletal muscle of the rat. Steroids, 28, 261-274.

Harper, A.E., Dwyer, J. and Brown, M.L. 1980. Human nutrition. In "Animal Agriculture: Research to Meet Human Needs in the 21st Century" (Ed. W.G. Pond, R.A. Merkel, L.D. McGilliard and V.J. Rhodes). (Westview Press, Boulder, Colorado). pp. 7-37.

Harte, F.J. 1969. Six years of bull beef production research in Ireland. In "Meat Production from Entire Male Animals" (Ed. D.N. Rhodes). (J. and A. Churchill Ltd., London) pp. 153-172.

Ravault, J.P. and Ortavant, R. 1977. Light control of prolactin secretion in sheep. Evidence for a photoinducible phase during a diurnal rhythm. Ann. Biol. Anim. Biochim. Biophys., 17, 459-473.

Schanbacher, B.D. 1982. Responses of ram lambs to active immunization against testosterone and luteinizing hormone-releasing hormone. Am. J. Physiol., 242, E201-E205.

Schanbacher, B.D. and Crouse, J.D. 1980. Growth and performance of growing-finishing lambs exposed to long or short photoperiods. J. Anim. Sci., 51, 943-948.

Schanbacher, B.D. and Crouse, J.D. 1981. Photoperiodic regulation of growth: a photosensitive phase during light-dark cycle. Am. J. Physiol., 241, E1-E5.

Schanbacher, B.D., Crouse, J.D. and Ferrell, C.L. 1980. Testosterone influences on growth, performance, carcass characteristics and composition of young market lambs. J. Anim. Sci., 51, 685-691.

Schanbacher, B.D, M.J. D'Occhio and T.W. Gettys. 1983. Pulsatile luteinizing hormone secretion in the castrate male bovine: Effects of testosterone or estradiol replacement therapy. J. Anim. Sci., 56, 132-138.

Schanbacher, B.D. and Ford, J.J. 1976. Luteinizing hormone, testosterone, growth and carcass responses to sexual alteration in the ram. J. Anim. Sci., 43, 638-643.

Schanbacher, B.D., Hahn, G.L. and Nienaber, J.A. 1982a. Photoperiodic influences on performance of market lambs. In "Proc. 2nd Int. Livestock Envir. Symp." (ASAE, Ames, Iowa) pp. 400-405.

Schanbacher, B.D., Hahn, G.L. and Nienaber, J.A. 1982b. Effects of contrasting photoperiods and temperatures on performance traits of confinement-reared ewe lambs. J. Anim. Sci., 55, 620-626.

Seideman, S.C., Cross, H.R., Oltjen, R.R. and Schanbacher, B.D. 1982. Utilization of the intact male for red meat production: A review. J. Anim. Sci., 55, 826-840.

Tucker, H.A. and Ringer, R.K. 1982. Controlled photoperiodic environments for food animals. Science, 216, 1381-1386.

DISCUSSION ON DR. SCHANBACHER'S PAPER

Dr. Heitzman: As regards the effect of intramuscular injection of trenbolone acetate on LH in mature heifers, it was not until all the measurable trenbolone had disappeared from the peripheral circulation that we began to get a return to LH episodic secretion. Once the trenbolone disappeared from the circulation the animal returned to normal.

Dr. Schanbacher: My intent is to look at the duration of implant effectiveness by relating it to other biological system effects such as effects on LH.

Dr. Heitzman: We have published a paper on the effects of trenbolone on oestrous cycle behaviour in mature dairy catle and LH was completely suppressed for anything between 3 and 6 months, and this coincided with presence of trenbolone in circulation.

Professor Karg: Why did you give a dose of 3 times the normal Synovex dose? In veal calves ½ the Synovex dose was shown to give better performance than a full dose.

Dr. Schanbacher: We did not. We gave the normal Synovex dose at three different intervals during the 205 day period. This was an attempt to reproduce the longer term effects expected from Compudose.

Dr. Galbraith: How long is the minimum pulse or flash of light given to your animals?

Dr. Schanbacher: One minute.

Dr. Galbraith: Is there a possibility that there is a stress effect on these animals? If for example you play music for 1 minute you will activate them. Would this elevate prolactin as seen after blood sampling or handling teats at milking?

Dr. Schanbacher: We entertained that question. During 24 hr. bleeds you learn what your animals are doing. It is difficult to evaluate stress or discomfort but the animals given a one minute flash as opposed to one hour flash do not even get up.

These animals have minimum access to move around. They do not look to be at all distressed or receiving anything other than retinal stimulation.

Dr. Forbes: As regards cortisol, it is lower with a 1 hr flash treatment. There is no evidence of any peak of cortisol at dawn, dusk or at the beginning, end or during the flash.

Dr. Buttery: Is there a bioreductase in muscle?

Dr. Schanbacher: I know there is in adipose tissue.

PHOTOPERIODIC EFFECT ON GROWTH AND FEED CON-
SUMPTION OF YOUNG BULLS

Martin Tang Sørensen
National Institute of Animal Science
Rolighedsvej 25, DK-1958 Copenhagen V

ABSTRACT

Extended photoperiod (16L:8D) had a negative effect on growth and feed consumption of bull calves during part of the initial . 10 weeks of treatment. After this period extended photoperiod stimulated weight gain and feed consumption. Feed conversion was not affected. Towards the end of a 24 week period with extended photoperiod the positive effect on weight gain decreased, while feed consumption remained relatively high. This trend continued after termination of light treatment and over the entire growth period to 360 kg live weight extended photoperiod had no effect on growth rate, a stimulating effect on feed consumption and a negative effect on feed conversion. There was no effect of extended photoperiod on body size, when animals were slaughtered at similar live weights.

INTRODUCTION

Peters et al (1978) found that heifers exposed to extended photoperiod gained weight 10% faster than control animals. On the other hand Roche and Boland (1980) found no effect of extended photoperiod on growth in bull calves. Their experimental period was only 84 days, however, and the results of Peters et al (1978) indicated that effects might be small during the initial weeks of extended photoperiod. On this background it was decided to conduct a "long-term" experiment to test, whether an effect similar to that in heifers could be found in young bulls. If so, it would be very beneficial for the beef industry. Especially if the increased weight gain rate was primarily due to an increased feed utilization, as the results of Peters et al (1978) indicated, and not to an increased feed consumption.

The objectives of the experiment, therefore, were to test whether young bulls exposed to extended photoperiod exhibited an increased growth rate, and if so, whether this was due to increased feed intake, and/or increased feed conversion. In addition the effect of light regime on carcass measurements was studied.

MATERIALS AND METHODS

The experiment was conducted at the Egtved test station, starting during the fall 1981.

Animals

290 bull calves were bought at commercial farms and brought to the Egtved station before they were 28 days old. The calves were offsprings of 29 sires from 3 breeds, i.e., Danish Black and White, Danish Red, and Danish Red and White. Each sire were represented by 10 offsprings. 36 animals were excluded from the analysis, because of failed pedigree test and health records.

Feeding

From arrival to 83 days of age the calves were fed restricted and equal amount of milk replacer, and from arrival to 98 days of age they were fed restricted and equal amounts of skim milk. Hay was ad libitum until 28 days, then from 28 to 56 days of age 0,2 kg per day were given and thereafter 0,3 kg per day. Until 84 days of age no barley straw was given, but 0,3 kg, 0,4 kg and 0,5 kg per day was offered from 84 to 112 days, from 112 to 168 days and after 168 days, respectively. Concentrates was fed ad libitum at all times.

Light

The animals were balanced with regard to sire and birth date and allocated to corresponding stalls in a building separated into two almost identical tie-stall barns. From October 1st to April 1st lights were on from 06.00 to 22.00 (16L:8D) in one barn, while in the other barn lights were on only when the herdsmen were working there, i.e. usually from 06.00 to 17.00 (normal light). In addition the animals were supervised a few minutes every evening, usually around 19.00. Between afternoon working hours and evening supervision there was a dim light not detectable by the lux-monitor in the barn with normal light.

In both barns the light intensity during dark hours with lights
on were between 30 to 40 lux at the level of the bulls' eyes.
At sunshine hours light intensity in both barns were >1000 lux.
The effect of light regime was completely confounded with barn.
However, analysis of previous data from the two barns could not
detect any "barn-effect". The temperature was about $14^{o}C$ in
both barns.

Recordings

The animals were weighed every 28 days, starting at the age
of 28 days. Feed consumption were also recorded in these 28-
day age periods. When the animals reached the weight of 360 kg,
they were slaughtered and carcass measurements were taken, i.
e. withers height, chest depth, heart girth, thurl width body
length and dressing percentage. The first animal was slaughtered April 1st, i.e., the same day as the light treatment terminated. Accumulated dry matter intake, growth rate and feed
conversion from age 28 to slaughter were calculated.

Statistical methods

Daily weight gain, daily dry matter intake and feed conversion (g weight gain per Scand.Feed Unit) in 28 day age periods were analysed according to the model

(1) $Y_{iljk} = \alpha + Sire_i + Light\ regime_l + Factor_j + Age\ period_k +$

$\qquad (Age\ period \times Light\ regime)_{kl} + b_k(W28_{ilj} - \overline{W}28) + \varepsilon_{iljk}$

$i = 1,...,29; \quad j = 1,...,5; \quad k = 1,...,7; \quad l = 1,2;$

The factor denotes an effect of vitamin E and selenium
supplement.

Weight at 28 days of age, $W28_{ilj}$ was included as a covariate. Only 7 age periods of 28 days were included, because of
reduction in number of animals due to slaughter.

The hypothesis H_o : LS means (1) = LS means (2) were
tested for the light regime and the age period x light regime
interaction.

Although the error variance was not quite constant for the

different age periods, an analysis on log transformed data showed that this did not disturb the results obtained from model
(1).

Accumulated performance data and carcass measurements
were analysed according to the model

(2) $Y_{ilj} = \alpha + Sire_i + Light\ regime_l + Factor_j + b_1(W28_{ilj}$
$- \bar{W}28) + b_2(W_{ilj} - \bar{W}) + \varepsilon_{ilj}$

Weight at 28 days of age and live weight at slaughter,
W_{ilj}, were included as covariates. The hypothesis H_o : LS means
(Normal light) = LS means (16L:8D) was tested.

RESULTS AND DISCUSSION

Some characteristics of the animals are shown in table 1.

TABLE 1   Characteristics of the 254 animals included
in the analysis.

Birth data	Weight 28 days of age (kg)		Age at slaught. (days)		Weight at slaught.(kg)	
$\bar{x}$ S.D.(day)	$\bar{x}$	S.D.	$\bar{x}$	S.D.	$\bar{x}$	S.D.
Sep.11th 20	51.1	6.6	287	29	361	6.0

It is seen that the "average birth date" was September
11th with a standard deviation of 20 days. This means that the
animals on the average were about 200 days old, when the light
treatment was terminated at April 1st. Blood samples were taken January 14th from a balanced group of animals from each
treatment for determination of serum prolactin. The samples,
however, have not been analysed as yet.

Table 2   Performance of young bulls exposed to two ligt regimes (leat square means from model 1).

Age period (days)	Sunrise to sun-set (h)	No.	Daily wt.gain (g)			Daily D.M.intake(kg)			Wt.gain/Scand.F.U.(g)		
			Norm	16L:8D	P	Norm	16L:8D	P	Norm	16L:8D	P
28-56	9.52	254	680	656	0.42	1.31	1.25	0.45	387	388	0.91
56-84	7.58	"	1019	954	0.03	2.14	2.02	0.08	377	365	0.14
84-112	6.57	"	1131	1125	0.84	2.96	2.89	0.30	360	366	0.44
112-140	7.38	"	1279	1368	<0.01	3.95	4.12	0.01	317	322	0.50
140-168	9.26	"	1337	1420	0.01	4.78	5.11	<0.01	269	264	0.52
168-196	11.34	"	1424	1482	0.06	5.67	6.04	<0.01	240	235	0.43
196-224	13.43	253	1483	1441	0.17	6.36	6.68	<0.01	223	205	0.02
28-224			1193	1206	0.26	3.88	4.02	<0.01	310	306	0.17

Table 2 and figures 1-4 show the performance of the bulls
in the two light regimes. Until about 100 days of age daily
gain and dry matter intake were lower of animals from 16L:8D
as compared to animals from normal light. Only in age period
56-84 was this effect significant, however, and an average effect from start of the experimental period to about the age
of 100 days would not be significant. This is in agreement
with results of Roche and Boland (1980), who found no effect
of extended photoperiod on gain and feed consumption during a
84-day period of bull calves of similar age.

From about 100 days of age to the end of the light treatment period at about 200 days of age, daily gain and feed consumption were highest on group 16L:8D. This result is in agreement with those of Peters et al (1978 and 1981) and Peticlerc
et al (1981) with older heifers. However, the 16L:8D bulls of
this experiment had no improved feed conversion as had the heifers in the above mentioned experiments.

Towards the end of the period with light treatment the differences in weight gain between groups decreased. This trend
continued after termination of light treatment, and the average
daily gain for the entire growth period was not different between the two groups (table 3). The bulls from the 16L:8D group
had the highest feed consumption for the entire growth period,
and thus must have maintained a relatively high feed consumption
after the end of the light treatment period. As a consequence
gross feed conversion was best for animals from the group with
normal light. The net feed conversion was highest for this
group too, because dressing percentage was the same for the

294

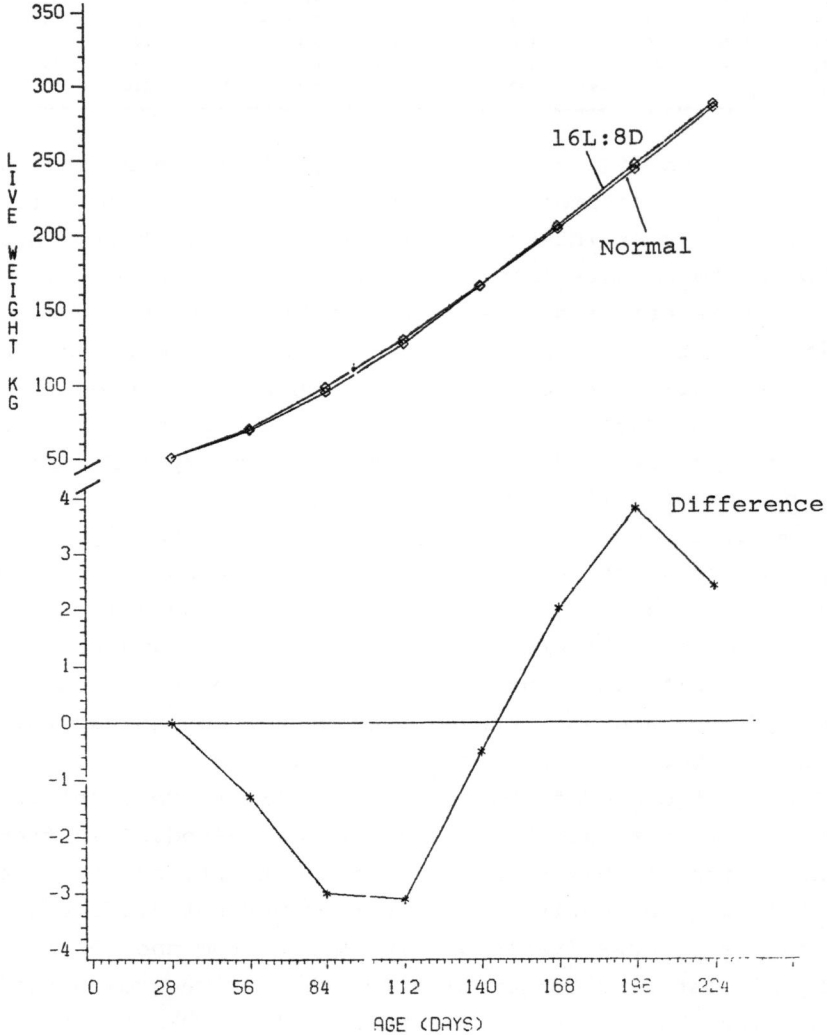

Fig. 1   Accumulated live weight of bulls exposed to two
light regimes and differences (16L:8D - Normal) due to
this treatment.

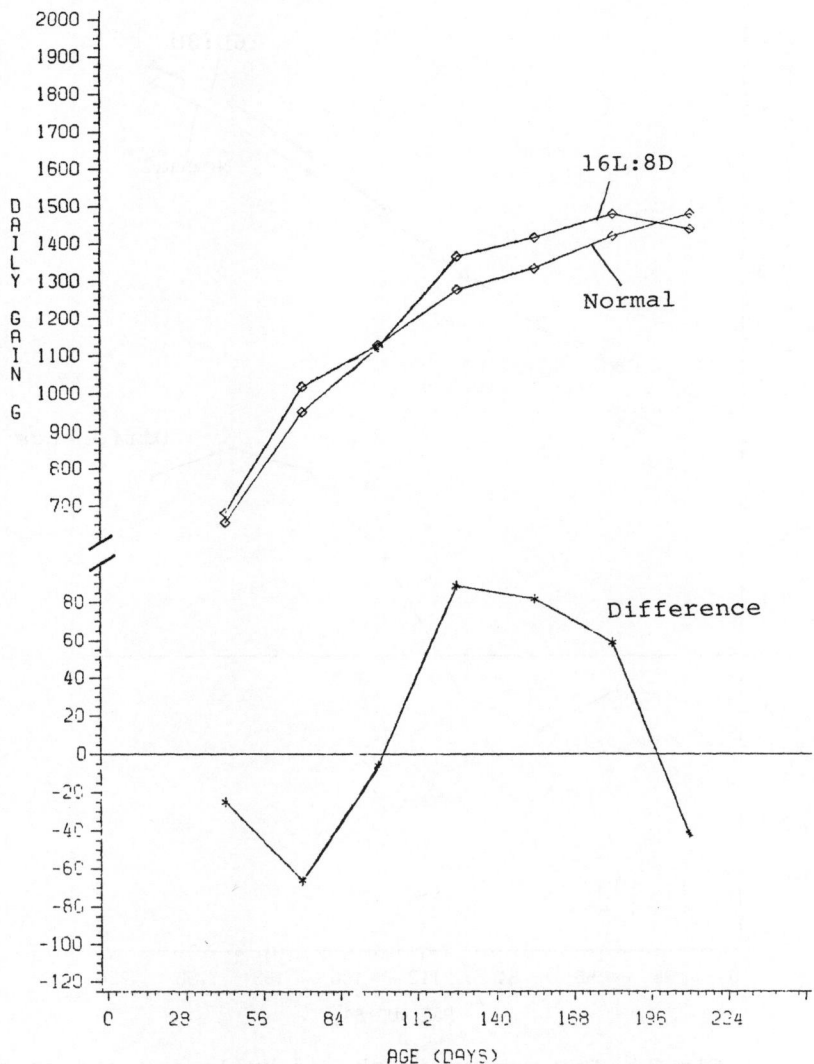

Fig. 2   Daily weight gain of bulls exposed to two light regimes and differences (16L:8D − Normal) due to this treatment.

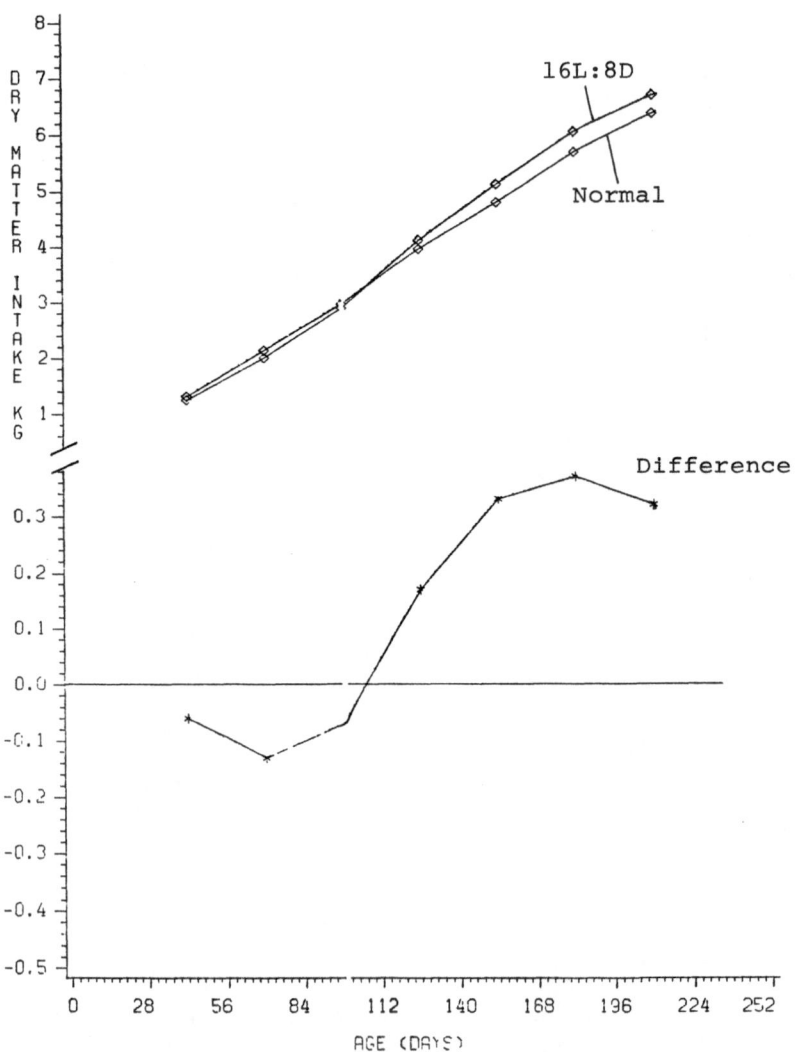

Fig.3   Dry matter intake of bulls exposed to two
light regimes and differences (16L:8D - Normal)
due to this treatment.

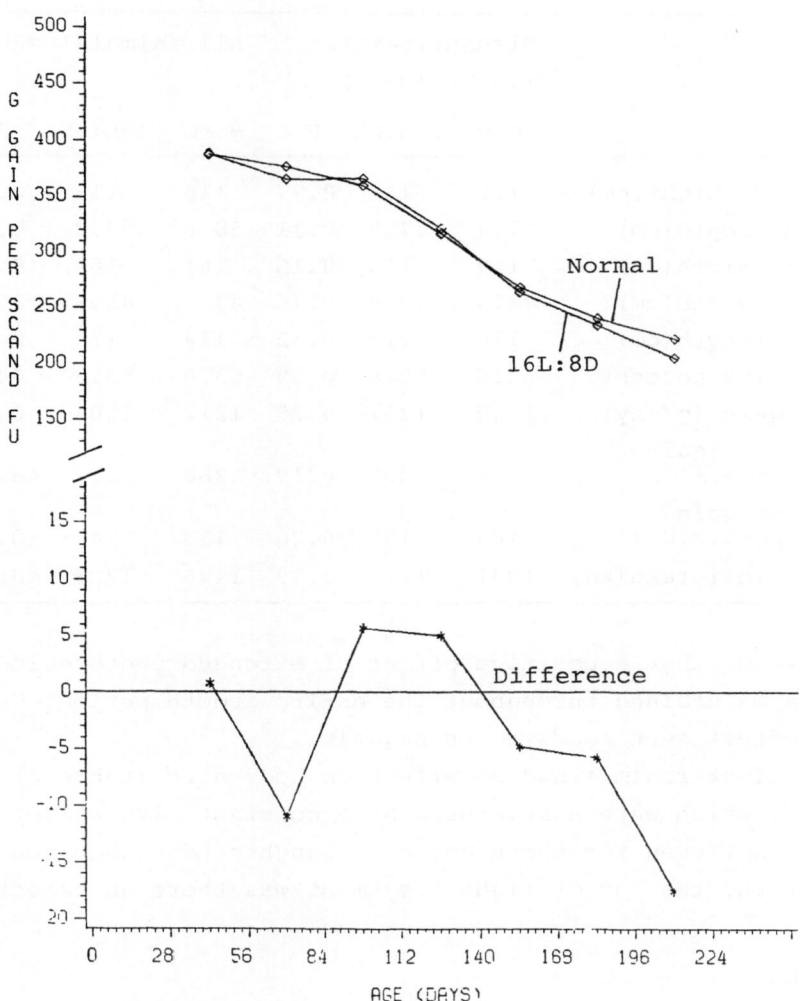

Fig. 4   Feed conversion of bulls exposed to two light
regimes and differences (16L:8D - Normal) due to this
treatment.

two groups.

TABLE 3   Carcass measurements and accumulated performance
data, i.e. from 28 days to slaughter, of young
bulls exposed to two light regimes (least sq.
means from model 2).

	Slaught.before 1.6.82 (N=65)			All animals(N=254)		
	Norm	16L:8D	P	Norm	16L:8D	P
Withers hight(cm)	116	116	0.77	116	116	0.64
Chest depth(cm)	57.4	57.9	0.11	58.6	58.8	0.16
Heart girth(cm)	159	160	0.15	161	161	0.52
Thurl width(cm)	42.9	42.8	0.76	43.3	43.1	0.12
Body length(cm)	114	113	0.32	114	114	0.49
Dressing percent	52.6	52.8	0.55	53.0	53.2	0.25
Ave.gain (g/day)	1339	1357	0.29	1217	1204	0.31
Gross wt.gain (g/Scand.F.U.)	309	302	0.19	288	277	<0.01
Net wt.gain (g/Scand.F.U.)	163	159	0.26	153	148	<0.01
Tot.D.M.intake(kg)	1091	1116	0.19	1176	1223	<0.01

It seems that a positive effect of extended photoperiod
cannot be maintained throughout the entire growth period. The
overall effect even seems to be negative.

The light regimes had no effect on body size (table 3) of
the bulls, which were slaughtered at a constant live weight
(360 kg), not even for those animals slaughtered within two
months of the the end of light treatment was there an effect.

REFERENCES

Peters, R.R., Chapin, L.T., Leining, K.B. and Tucker, H.A.1978.
   Supplemental lighting stimulates growth and lactation
   in cattle. Science 199:911-912.
Peters, R.R., Chapin, L.T., Emery, R.S. and Tucker, H.A. 1981.
   Growth and hormonal response of heifers to various photo-
   periods. J.Anim.Sci. 51:1148-1153.

299

Peticlerc, D., Chapin, L.T., Emery, R.S. and Tucker, H.A. 1981.
    Interaction of photoperiods and planes of nutrition on
    growth, puberty and growth hormone in Holstein heifers.
    J.Anim.Sci.53 (suppl.1): 357-358.
Roche, J.F. and Boland, M.P. 1980. Effect of extended photo-
    period in winter on growth rate of Friesian male cattle.
    Ir.J.Agric.Res. 19:85-90.

DISCUSSION ON DR. SORENSEN'S PAPER

Dr. Schanbacher: Michigan work shows a positive effect of photoperiod on milk yield. With regard to chest size in heifers there is a prolactin response that might not be the case in steers. Do you think that perhaps we have not tested adequately light quality or intensity in order to come to the conclusion that we should be using supplemented light in cattle?

Dr. Sorensen: The Michigan work on growth only includes data from the experimental period. The data terminates when the light treatment terminates. This work shows that if the period after light treatment termination is included, the overall effect on growth is not significant. There is not really any discrepancy between this work and the Michigan work.

Dr. Roche: In Ireland, supplementing normal day light with extra light does not increase growth rate in cattle based on three years of experiments using calves, pre-pubertal heifers or bulls, steers and adult heifers. Different intensities of 100 to 600 lux were used.

Dr. Robertson: Did you notice any differences in behaviour in the animals, especially in the initial stages?

Dr. Sorensen: We did not take any particular measurements but I was not aware of any differences.

GENERAL DISCUSSION

<u>Dr. Roche</u>: I would like to discuss some result of an experiment we have carried out at Grange with Vivian Reynolds. One difference between the Michigan State work and our work was the use of male castrates at Grange. We carried out an experiment in 2 year and 1 year old heifers to see if we could repeat the Michigan State work, which was done in heifers. Supplementing normal light from 16.00 hr. to 24.00 hr or giving extra light from 24.00 hr. to 08.00 did not significantly increase growth or carcass weight of heifers. Of the twelve one year old heifers that were slaughtered, five had ovulated and therefore had reached puberty, where as none of the control heifers had reached puberty. From this experiment, we have concluded that extra light in 1 or 2 year old heifers has no significant effect on daily liveweight gain or carcass weight but there may be a stimulatory effect on puberty in 1 year old heifers. All these animals were being given supplemental lighting in addition to natural daylight. We followed this experiment by looking at intact and castrate males and females, 1 yr. old. The only effect on growth rate in the experiment was a decrease due to castration of the males. There was no effect of extra light on daily liveweight gain. We are happy to conclude that we are not getting any beneficial response from supplementing normal light for eight hours either at 16.00 or 24.00 hrs. under Irish conditions.

<u>Dr. Schanbacher</u>: We did a similar trial with cattle. We gave two hours supplemented light between 2200 hours and 2400 hours in bulls and steers, and we did not find an increase in weight gains in either of those groups. Prolactin might be a means of looking at what happens when you use artificial light in addition to natural light. However, the qualities are so different, perhaps in doing 24 hour profiles of prolactin we could determine if supplemental artificial light affects prolactin when given in addition to natural day light.

Dr. Forbes: In our experiments in which natural light was extended with artificial light, we did not measure prolactin to see if it was responding, even though growth was not.

Dr. Roche: We got variable responses to prolactin. Certainly there is increased prolactin release to TRH challenge in the light treated animals in comparison to controls. It does not look as if there was an increase in prolactin from light treatment.

Dr. Schanbacher: Dr. Tucker suggests that you will not pick up differences in prolactin in steers during the winter because levels are already so low.

Dr. Forbes: They have shown that in winter where temperatures were around zero prolactin was very low and there was no difference due to photoperiod, but there still was a significant effect on liveweight gain.

Dr. Boland: We carried out a trial comparing the effects of different photoperiods or anabolic agents on growth rate of store lambs with thirty animals per treatment. The average daily liveweight gain for the 16L:8D, 8L:16D, natural photoperiod and oestradiol (7.5mg) + trenbolone acetate (80 mg) + progesterone (75mg) were 149 ± 13, 128 ± 14, 141 ± 14 and 183 ± 14 respectively. There was no significant difference ($P>0.05$) between the different photoperiods, but a significantly higher growth rate was obtained with the anabolics ($P<0.05$). Lambs exposed to 16L:8D grew slightly faster than those exposed to 8L:16D, but at a comparable rate to those exposed to natural photoperiod. A better interpretation may be that 8L:16D depressed growth rate to some extent. It may be very relevant to have a group of animals on natural photoperiod in further trials to determine the effects of photoperiod.

Dr. Forbes: A flash of light is in contrast to the dark of night. It is not like leaving lights on to try and extend natural light.

Dr. Roche: Yes, but you should compare growth rate in animals under natural light conditions and not under 8 hours of light. It is possible that 8 hr. light decreased growth rate and 16 hr. gives similar growth to controls.

<u>Dr. Schanbacher</u>: We have lambs gaining 600 grams per day, on experiment. If you compare these with controls in an outside environment, the gains would be about 300 grams. The environment of the experimental animals is involved in making those animals gain better. We would not find our short day animals doing as poorly as that of the controls.

<u>Dr. Roche</u>: The winter conditions in your area are more severe than in Ireland. We have not got one significant response in cattle or sheep to supplemental or controlled light. I might add that in two trials where we attempted to increase milk yield, we got no effect from giving supplemental light in winter to the dairy cow, either.

<u>Dr. Quirke</u>: We conducted an experiment where sheep were induced to lamb in March, September or November. The ewes were fed for the first 6 weeks of lactation. The lambs born in November were about 20% lighter at six weeks of age than lambs born at other times of the year suggesting that shorter day length had a depressing effect on lamb growth rate.
Question: Did you measure milk yield?
<u>Dr. Quirke</u>: We would assume it was controlled by feed intake which we were controlling.

<u>Dr. Boland</u>: Were they all housed indoors?

<u>Dr. Quirke</u>: They were all housed in an open plan shed.

<u>Dr. Schanbacher</u>: Dr. Terqui commented about the importance that a light flash should be placed 9 hours after dusk as opposed to our 17 hours after dawn. Where we are imposing light flashes in cattle and sheep under natural environment we are likely not to be positioning the light flash at the correct time at all, and we should be positioning the light flash not at hour 17 or at midnight but we should make sure we have the light flash appropriately positioned.

<u>Dr. Roche</u>: In one of our experiments with cattle, we gave

supplemental light from 16.00 to midnight, and another from midnight to 08.00. Nine hours after dusk would be 01.00 hr. That did not work. In another group we gave them extra light from 22.00 to 02.00 hr. That did not work either.

<u>Dr. Schanbacher</u>: It is likely that a light flash for an hour at the right time is not the same thing as light coming on at that same time and then continuing until the next light phase i.e. until morning. It might be helpful to differentiate the location of the animal or the season of the year.

<u>Dr. Terqui</u>: The photosensitive phase is not fixed. The photosensitive phase is moving with time, according to the season.

SUMMARY AND CONCLUSIONS

The CEC under the auspices of the Beef Research Programme organized a meeting on "Growth Control in Cattle" which was held at CEC Brussels Dec. 13th and 14th. Scientists from 9 member states, the U.S. and representatives from certain pharmaceutical companies attended. The meeting endorsed the conclusions of the Expert Scientific Group on Anabolic Agents. Recent research on metabolism, pharmacokinetics and problems of fixing tolerance levels of xenobiotic anabolic agents were discussed. The present position on analytical techniques was updated with particular reference to xenobiotic anabolic agents.

Methods to modify the sexual behaviour of bulls were discussed and data presented suggested the possibility of reducing aggressive behaviour and improving growth rate at the same time. Meat quality of veal calves given anabolic agents was shown not to be different to that of untreated veal calves. The possibility of increasing growth rate by immunization of lambs against somatostatin shows initial promise and demonstrates the importance of further work in growth physiology as a method of increasing efficiency of animal production.

The present state of endocrinology of growth and the mechanism of action of anabolic agents is characterized by a need for more knowledge and research in this area. Discussion on increasing growth by manipulation of the length of the photoperiod indicated some positive results in sheep but generally no response in cattle.

The report of the Expert Scientific Group on Anabolic Agents was discussed. There was no divergence of opinion between the two groups and the following recommendations were made:

(a) The use of oestradiol - 17β, testosterone and progesterone and those derivatives which readily yield the parent compound on hydrolysis after absorption from the site of application, would not present any harmful effects to the health of the consumer when used under the appropriate conditions as growth promoters in farm animals.

(b)   Evaluation of the data on "trenbolone" and "zeranol" revealed that some data on the hormonal non-effect level and the toxicology of these compounds and their metabolites are lacking. Additional information is required before a final conclusion can be given on trenbolone and zeranol.

(c)   Proper programmes to control and monitor the use of anabolic agents are essential.

(d)   It is necessary to continue scientific investigations on the relevance of the present use of the "no-hormone effect" when related to the harmful effects of anabolic agents.

Priority areas for further research were identified as follows:

(i)   Development of simple, fast but reliable analytical technieques for residue analysis.

(ii)   The physiology of growth in beef cattle.

(iii) Mode of action of growth promoters.

(iv)   Development of methods, other than the administration of anabolic agents, to increase growth, carcass weight and quality viz. extended photoperiod, manipulation of the immune system, practical administration of growth hormone or somatomedins, and manipulation of rumen function.

(v)   Cellular aspects of muscle biology.

(vi)   Multidisciplinary approach to the problem is required.